GREAT
BREAKTHROUGHS IN
MATHEMATICS

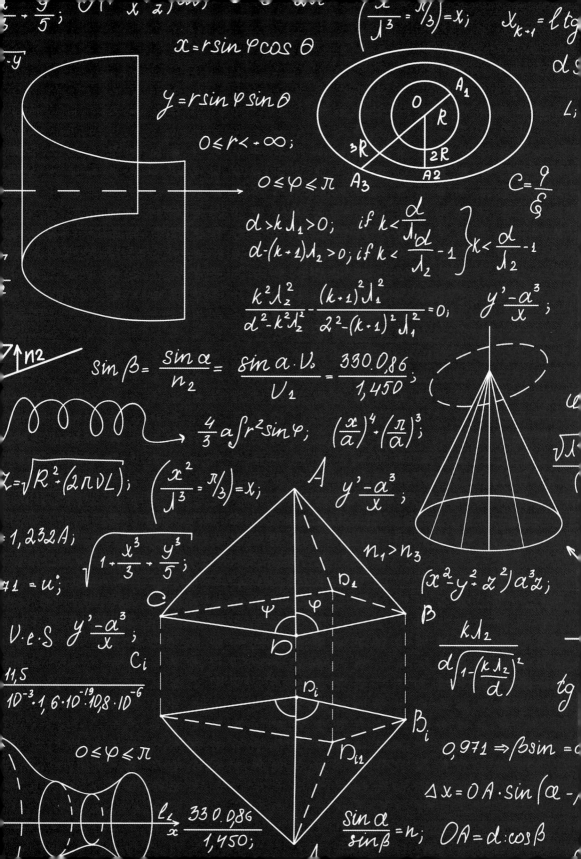

GREAT BREAKTHROUGHS IN MATHEMATICS

FROM COUNTING TO CHAOS THEORY – HOW NUMBERS CHANGED THE WORLD

ROBERT SNEDDEN

SIRIUS

Picture credits

All efforts to contact copyright holders have been made by the publisher. Any omissions will be rectified in future editions.

Bridgeman Images: 16
Flickr: 19
Getty Images: 22, 71, 75, 83, 95, 140, 162, 185, 188
Science Photo Library: 17, 33, 148
Shutterstock: 2, 7, 8, 9, 11, 13, 14, 15, 20, 23, 27, 31, 32, 36, 39, 40, 42, 44, 44, 46, 47, 48, 50, 53, 55, 56, 68, 70, 70, 72, 77, 81, 87, 89, 90, 94, 96, 98, 99, 100, 103, 106, 107, 108, 113, 115, 116, 123, 124, 135, 136, 140, 141, 147, 150, 156, 156, 157, 158, 160, 161, 165, 167, 169, 171, 173, 176, 178, 183, 187, 188
Pixabay: 38
Wikimedia Commons: 10, 12, 23, 24, 26, 29, 30, 34, 43, 45, 48, 58, 59, 60, 61, 63, 64, 66, 78, 80, 86, 93, 97, 97, 104, 110, 114, 118, 119, 120, 121, 122, 126, 127, 127, 129, 130, 131, 134, 138, 138, 141, 142, 144, 146, 146, 153, 154, 168, 172, 174, 180, 180, 181, 182, 182, 184, 186

All other diagrams and artwork were created for this edition by David Woodroffe.

SIRIUS

This edition published in 2020 by Sirius Publishing, a division of Arcturus Publishing Limited,
26/27 Bickels Yard, 151–153 Bermondsey Street,
London SE1 3HA

ISBN: 978-1-83940-684-3
AD005933UK

Printed in Singapore

Contents

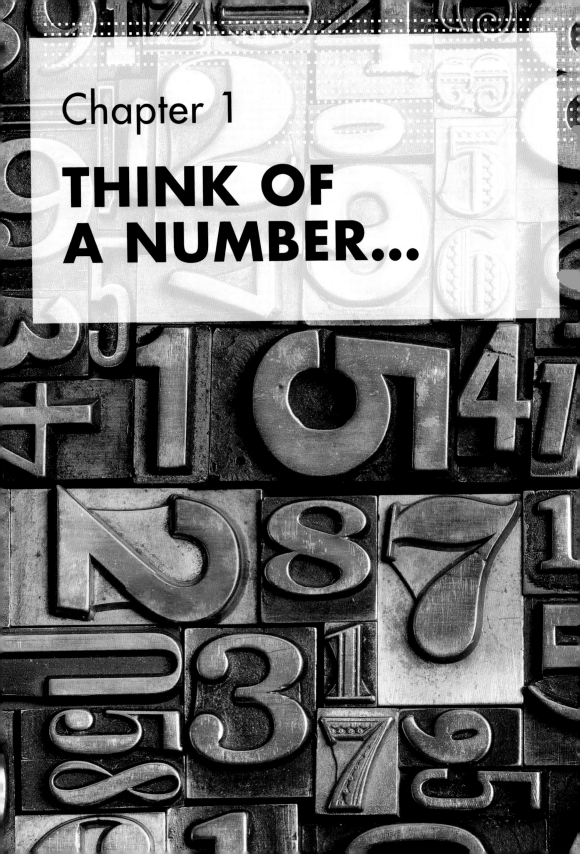

Chapter 1

THINK OF A NUMBER...

Think of a Number...

'Mathematics is the art of problem solving.'
George Pólya, mathematician.

THE PROBLEM OF QUANTITY

Getting through life requires the ability to access information in a variety of forms about all manner of things. One type of information we use a lot comes in the form of numbers. How much money do I have in the bank? How many minutes do I have to wait until my train arrives? How many marks did I score in the test? But it hasn't always been this way.

Our hunter-gatherer ancestors had to remember which fruits were poisonous and which were good to eat; they needed to know where animals could be found and how they were likely to behave. But almost certainly they didn't count the number of berries on a bush or the number of animals in a herd. So, when did people first start using numbers? When did knowing how much there was of something become a problem?

Cave painting from Magura Cave in Bulgaria.

Number timeline	
*c.*37,000BC	Notched bones may provide the first evidence that humans are thinking about numbers.
*c.*4,000BC	Evidence of the use of numbers begins to appear in the river valley civilizations of the Indus, Nile, Tigris and Euphrates, and Yangtze.

READING THE BONES

The language of mathematics is expressed in numbers. If we hadn't learned to count and use numbers, civilization as we know it would have been impossible. It has been suggested that numbers and mathematics are so fundamental to our nature that mathematical thinking must have grown in parallel with the development of human thinking processes in general.

Number awareness

Can you be aware of a number without having a name for it? Present-day hunter gatherers such as the Warlpiri in Australia count 'one, two, many', while the slightly more numerate Munduruku of South America have no names for numbers greater than five. If the Warlpiri only have words for one and two could they, for example, choose between four pieces of fruit or five? Neuroscientist Brian Butterworth carried out an experiment in 2008 in which he asked Warlpiri children to lay out the number of counters that corresponded to the number of sounds he made. The Warlpiri did just as well as children from English-speaking backgrounds. Though they lacked the language to express it, the Warlpiri were just as aware of quantity as anyone else.

Aboriginal rock paintings in Kakadu National Park, Australia. The Warlpiri are a numerate people, however they do not express their numeracy in linguistically similar ways to Western cultures.

We cannot know how early humans viewed the world. All we can do is to make guesses based on the artefacts they left behind. The earliest physical evidence we have to suggest mathematical thinking is the Lebombo bone. This fragment of a baboon's leg bone was found in a cave in the Lebombo mountains of Swaziland in southern Africa and is approximately 37,000 years old. Carved into the surface of the bone are 29 distinct notches. The exact purpose of the bone is unclear. But it has similarities with the calendar sticks still used by present-day hunter gatherers in Namibia, and it has been suggested that it was a tool for keeping track of the days of the lunar month or of a woman's menstrual cycle. Were African women the world's first mathematicians?

In 1937, a wolf bone was unearthed in what is now the Czech Republic. Estimated to date back to 30,000BC, the bone is engraved with 55 deeply cut notches in groups of five, which is the traditional grouping used for tallying. (See the box 'Subitize this' for why this may be so.)

THE ISHANGO BONE

Another famous example of a possibly mathematical artefact are the two Ishango bones. The first, and most famous, was discovered in the 1950s at Ishango village in what is now the Democratic Republic of Congo. It has been dated at around 22,000 years old. The Ishango bone's carvings are more complex than those of the earlier Lebombo bone. There are three sets of grooves, carved in sequences of:

19, 17, 13, 11

7, 5, 5, 10, 8, 4, 6, 3

9, 19, 21, 11

What could be the significance of these number groups? The bone's discoverer, Belgian geologist Jean de Heinzelin de Braucourt, suggested that the notches might represent a game based on arithmetic and that the patterns formed by the notches indicated that the carver used a base-10 counting system and had knowledge of multiplication. Others have suggested that, like the Lebombo bone, the Ishango bone was also a timekeeper, used to keep track of the phases of the moon.

It's worth keeping in mind that we have no idea of the context

The first Ishango bone, which is estimated to be about 22,000 years old.

in which these bone markings were made – we are only inferring a purpose for bones such as these from our present-day standpoint. The Lebombo bone is broken so there may well have been more than 29 incisions on it, which would put an end to speculation that it was a lunar calendar reckoner. And even if 29 incisions were all there were, their purpose could still be something quite unknown to us. It might have been used by someone checking the sharpness of recently made flint tools, for example. The grooves on the Ishango bone might be there to make it easier to grip. Or perhaps the carver was simply idling away the time by the campfire as he, or she, waited for the Sun to rise on another day.

Subitize this

Subitizing is the ability to look at a small number of objects and see how many there are without actually counting them. We do it all the time. If you take four or five coins out of your pocket you can see at a glance how many there are without needing to count them. It does appear, though, that five is the maximum set of objects our brains can recognize – beyond that, we do have to count. This 'number sense' is an ability that isn't particular to humans – honeybees, birds and monkeys have all demonstrated that they can do the trick – but only humans took the next step and began to count things that lay beyond the scope of subitizing.

When you can see, just at a glance, how many coins there are without having to count each one, this is known as subitizing and it's a good example of our ability for 'number sensing' in action.

TALLY HO!

We have no idea when numbers and counting became important, but it is likely that it began about 10,000 years ago, around the time that people began to settle in one place and farm, rather than hunt and gather. It might not have mattered how many wild boar there were in a herd before you started hunting them, but if you started the day with 20 sheep in your flock you'd probably want to know that you still had 20 sheep at the end of it.

A shepherd with 20 sheep could tally them off against grooves cut in a tally stick, or indeed by using his or her fingers and toes. Another way is to make a small heap of pebbles, each pebble representing one sheep in a simple one-to-one correspondence. There's no need even to count them – all that need be done is to remove one pebble from the pile every time a sheep is returned to its pen in the evening. In fact, the word 'calculation' has its root in the Latin *calculus*, which means pebble.

The thing about pebble counting, or any other form of tallying, is that it doesn't actually require numbers – just a correspondence between one physical object and another. But as civilization took root and life grew more complex, the need for numbers to keep track of everything became inevitable.

Tally sticks from the Swiss Alps.

Chapter 2

NUMBER SYSTEMS

Number systems

As settlements grew larger, accumulating people, goods and livestock in greater and greater numbers, a way had to be found to keep track of it all. To solve the problem of how many of something you have you need a number system. Different cultures in different times and places have developed a variety of solutions to this problem.

An ancient Egyptian carving from the reign of Mentuhotep II, c.2051–2030BC showing hieroglyphics.

Number systems timeline

*c.*3,400BC	The Egyptians use simple straight lines as the first symbols for numbers.
*c.*3,000BC	A decimal number system is in use in Egypt.
*c.*3,000BC	The Egyptians develop a hieroglyphic system of numerals.
*c.*3,000BC	The Babylonians use a sexagesimal number system for financial transactions; although it is a place-value system it has no zero-place value.

THE MATHEMATICIANS OF MESOPOTAMIA

The Sumerians, one of the earliest of the Mesopotamian civilizations, were possibly the first to develop a system of numbers and counting around 4,000BC. Their base unit was 60, a system that persists to the present day, for example, in the division of a minute into 60 seconds, an hour into 60 minutes and a circle into 360 degrees. Later civilizations, such as the Egyptians, would use the base-10 system that is more familiar to us today.

So impressive are the achievements of the people of Sumer (a region of Mesopotamia, now part of modern-day Iraq) that it has been described as the 'Cradle of Civilization'. The innovations of the Sumerians can be said to have shaped all the civilizations that came after them and include the wheel, agriculture, irrigation and much more. The Sumerians developed the earliest-known writing system, cuneiform script, which used wedge-shaped characters inscribed on baked clay tablets. The long-lasting nature of these clay tablets has led to us having more knowledge of the mathematics of ancient Sumer and Babylon than of early Egyptian mathematicians, whose work was recorded on the more perishable papyrus.

Ancient Sumerian stone carving with cuneiform script.

COUNTING CLAY

It is probably impossible to have a civilization without any form of bureaucracy to keep it in order and Sumerian mathematics initially developed to meet the needs of its bureaucrats.

They were, perhaps, the first people to move away from using tokens of some sort to represent sheep, jars of oil and other commodities and begin using number symbols to indicate quantities. By about 3,000BC, the Sumerians were drawing images of tokens

Example of Sumerian tokens used for counting.

on clay tablets. Different types of goods were represented by different symbols, and multiple quantities represented by simply repeating the symbols. The drawbacks to this system are fairly obvious. Everything has to have its own sign, each of which would have to be learned. Also, while it works well for small numbers, having to make tally marks for 300 sheaves of wheat, say, would be time consuming and prone to error.

A major step forward came with the introduction of symbols to denote quantity. These were distinct from the symbols for goods. Rather than show ten oil-jar symbols, there would be one oil-jar symbol plus the symbol for the number '10'. A system of this sort is known as a 'metrological numeration system' – it is really a system of weights and measures. The number symbol is given context by being attached to the goods symbol and isn't really thought of as an abstract idea in itself as numbers would later come to be seen.

$\sf D$	$\xrightarrow{.10}$	$\sf O$	$\xrightarrow{.6}$	$\sf D$	$\xrightarrow{.10}$	$\sf D$	$\xrightarrow{.6}$	$\sf O$	$\xrightarrow{.10}$	$\circledcirc$
= 1		= 10		= 60		= 600		= 3 600		=36 000

The number system used by Sumerians before cuneiform symbols were used to mark numbers.

In the base-60 sexagesimal system used for counting, a single object was indicated by a small cone. Ten cones equalled one small circle, six small circles equalled one big cone, ten big cones equalled a big cone with a circle inside, six of those equalled a large circle and ten large circles equalled a large circle with a small circle inside. This means the last unit was worth 10 × 6 × 10 × 6 × 10 = 36,000 base units. Over the course of the next few centuries, these signs were gradually replaced by cuneiform equivalents inscribed on clay tablets

'A mind that does not know accounting, is it a mind that has intelligence?'

Ancient Mesopotamian proverb.

PLACE-VALUE NOTATION

The invention of place-value, or positional, notation by the mathematicians of Babylon was a major innovation in the use of numbers. It meant that the value of a number was indicated not only by its symbol but also by its position. Consider the number 333 – all three symbols are identical, but one '3' stands for three hundreds, one '3' for three tens and one '3' for three units.

In terms of their number system, the Babylonians inherited some ideas from the Sumerians and from the Akkadians, such as the base 60 sexagesimal system. Neither the Sumerian nor the Akkadian system was a positional system, however, and this advance by the Babylonians may be ranked as their greatest achievement in the development of their number system. The advantage, and indeed necessity, of a place-value system would have become obvious as civilization grew more complex and faced the problem of having to deal with ever larger numbers.

The fact that the Babylonians used a base-60 counting system might at first lead you to believe that they must have had a lot of symbols to learn, but in fact they used just two – a unit symbol and a ten symbol – so '6' was represented by a group of six unit symbols and '26' was indicated by two ten symbols and six unit symbols. Just as in our decimal system, Babylonian numbers read from left to right, so the rightmost position was for numbers

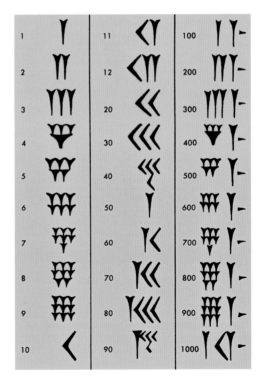

The Sumerian sexagesimal numbers.

up to 59 and positions to the left of that for increasing powers of 60.

A potential and obvious problem with the system is that '2' is represented by two one-unit characters, but so too is '61', although in the latter number one of the units is actually in the second place and stands for 60. Nonetheless, the Babylonian numbers for '2' and '61' look pretty much the same. The Babylonians solved this problem by having the units for '2' touch each other to become essentially a single symbol, while leaving a space between the characters for '61'. Still, it is easy to see how a careless scribe could cause some serious errors in accounting!

A much bigger problem was the lack of a zero to put into an empty position. The sexagesimal numbers for '1' and '60' looked exactly the same, for example: one was just supposedly a little further over to the left than the other. However, the Babylonians appeared to have coped very well with their system as it was, and doubtless context made it clear what the value of the symbol was. Eventually, they did invent a symbol to indicate an empty space.

Why base 60?

Why did the Mesopotamians base their number system on 60? There are several suggestions, none of them entirely satisfying. Theon of Alexandria tried to answer the question in the 4th century AD, suggesting that 60 was chosen because so many numbers divide evenly into it (2, 3, 4, 5, 6, 10, 12, 15, 20, 30) making it easier to work with fractions.

Based on the weights and measures used by the Sumerians, Austrian mathematician Otto Neugebauer suggested that a decimal counting system was modified to base 60 to allow for dividing weights and measures into thirds. He may be correct, but it's just as likely that the system of weights and measures adopted was dictated by the number system rather than the other way around.

Other theories, some of them far-fetched, have been based on astronomy. Mathematics historian Moritz Cantor suggested that choosing 60 came from dividing up a year of 360 days, but the theory doesn't really stand up as the Sumerians were certainly aware that the year was longer than 360 days. The suggestion that 60 is arrived at by multiplying the number of moon cycles per year with the number of naked-eye planets (Mercury, Venus, Mars, Jupiter, Saturn) again seems highly unlikely and arbitrary as a basis for adopting a numbering system.

There are also theories based on geometry. One theory proposes that the Sumerians regarded the equilateral triangle as the fundamental geometrical building block. The angles of an equilateral triangle are 60° and if this was divided by 10, to give a basic angular unit of 6°, there would be sixty of these basic units in a circle. It's not wholly convincing!

Chapter 3

THE BEGINNINGS OF GEOMETRY

The beginnings of geometry

The word 'geometry' comes from the Greek words for Earth (*geo*) and measurement (*metry*). It is the branch of mathematics that deals with lines, shapes and spaces and their relationships with each other. Geometric principles were discovered independently by various cultures around the world as settled societies faced the problem of accurately dividing up their resources for purposes of ownership and taxation.

The ancient Egyptians developed a number of sophisticated techniques for calculating areas and volumes, but they took a pragmatic, 'rule of thumb' approach to the subject and didn't advance any theories of geometry. The Babylonians were also adept mathematicians who solved many geometry problems. The Greeks recognized the achievements of the Egyptians and Babylonians but sought to place geometry on a sound basis of proof and reasoning, an enterprise that was driven forward by renowned mathematicians of the ancient world such as Thales, Euclid and Pythagoras.

Statue of Euclid, the great ancient Greek mathematician and founder of geometry.

Geometry: Solving problems over time

C.5,000BC Early Egyptians and Sumerians use geometric designs but this is more art than mathematics.

C.3,000BC There is evidence that megalithic societies in Northern Europe had a good grasp of geometric principles.

C.1,550BC The Egyptian scribe Ahmes transcribed a text that was written by an unknown hand some 200 years earlier. The Ahmes papyrus set out more than 80 mathematical problems, complete with solutions, including how to calculate the volume of a granary.

C.575BC Greek mathematician and philosopher Thales brought his knowledge of Egyptian and Babylonian mathematics to Greece. He used geometry to solve problems such as calculating the height of pyramids and the distance of ships from the shore.

C.300BC Euclid of Alexandria organized existing knowledge of geometry in his *Elements*, one of the most famous and influential of all maths texts. Apollonius later wrote *Conics*, in which he introduced the terms 'ellipse', 'parabola' and 'hyperbola'.

C.140BC Hipparchus began the development of trigonometry.

The Rhind Papyrus

The Rhind Papyrus, named after the Scottish antiquarian Henry Rhind, who bought it on a trip to Luxor in 1858, offers a fascinating glimpse into the world of Egyptian mathematical thinking. It is also known, perhaps more fairly, as the Ahmes Papyrus, after the scribe who copied it between 1650 and 1500BC from a text that was written around two centuries earlier.

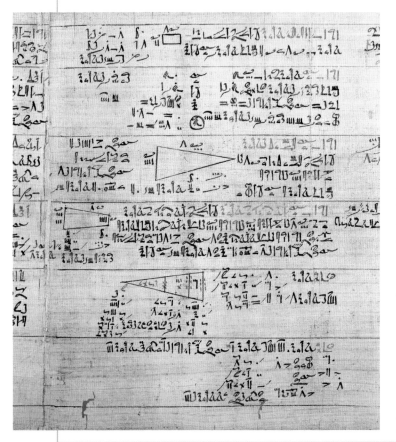

According to Ahmes, the papyrus provided, 'Accurate reckoning for inquiring into things, and the knowledge of all things, mysteries ... all secrets.' It contains reference tables to aid in calculations and sets out more than 80 mathematical problems of the sort that would be familiar to the Egyptian civil servants of the time, such as calculating the volume of a granary.

A fragment of the Rhind Papyrus.

The ability to measure areas accurately was very important to the Egyptians' agricultural civilization. Land was subject to tax, so both tax inspectors and land owners had an interest in knowing that their liability was being correctly calculated. When a parent died it was traditional to divide up the land between the children; again, accurate measurement was essential to avoid dispute. The surveyors of Egypt also had the job of redrawing property boundaries after the annual flooding of the river Nile washed away the markings distinguishing one field from another.

Five of the problems in the Ahmes Papyrus are specifically concerned with pyramids. A particular problem involves the calculation of the slope of the side of a pyramid, a vital concern for the builders, who had to make sure that the slope on all four sides of the pyramid was uniform.

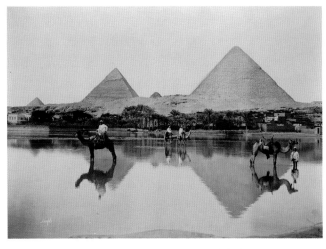

A photograph showing an Egyptian village during the Nile flood circa 1900.

Egypt's surveyors were known as the *harpedonaptai*, or rope-stretchers, as the principal tool of their trade was a length of knotted rope that they used to measure out property boundaries. Each knot was spaced at intervals of a cubit. The cubit was originally the distance from the elbow to the tip of the middle finger but, as this could vary from person to person, it was standardized as the royal cubit at 52.3 cm, as evidenced by examples of cubit rods that have survived.

The *harpedonaptai* could do more than measure straight lines with their ropes. They knew that a rope divided into 12 equal sections could be used to form a triangle with sides of 3, 4 and 5 units, which yields a perfect right angle. The Egyptian surveyors used this knowledge when laying the foundations for their buildings, including the pyramids. It seems most likely that the Egyptians discovered this relationship between the length of a triangle's sides and the forming of a right angle by accident. There is no evidence that they formulated a theorem to explain it. That would have to wait for Pythagoras.

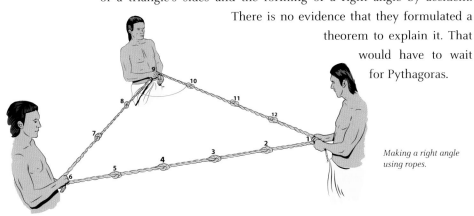

Making a right angle using ropes.

Pythagorean theorem

Pythagoras' theorem is probably high up on the list of what most people remember from their mathematics classes.

Its simple formula, $a^2 + b^2 = c^2$, makes it easy to recall. It tells us straightforwardly that the square of the length of the longest side, the hypotenuse (c) is equal to the sum of the squares of the two shorter sides (a and b) of a right-angled triangle. Although the principle was known for several centuries before Pythagoras, we name the theorem after him as he was the first to provide a proof for it. The 3, 4 and 5 rope lengths of the *harpedonaptai* form the first of the 'Pythagorean triples', an infinite set of whole numbers that are solutions to the theorem. Other examples of Pythagorean triples are, 5, 12, 13; 7, 24, 25; and 29, 420, 421.

All square

It was the Egyptians who came up with the notion of determining an area in terms of the number of squares of one standard unit measurement of length on each side that could be fitted into it. In their case, the standard unit was the cubit; the more cubit squares of land you had, the greater your tax liability. We use the same system today, though we would use metres or feet, for example, as in saying a garden has an area of 15 square metres. In working out how many tiles we need to cover the bathroom wall we're tackling a similar problem to the one solved by the *harpedonaptai*.

Measuring an area that is square or rectangular is straightforward. The task becomes trickier if it is triangular, say, or circular. The geometers of ancient Egypt discovered the formula for calculating the area of a triangle – $\frac{1}{2} \times b$ (the base length) $\times c$ (the height) – and also knew how to calculate the area of a quadrilateral.

One of the problems in the

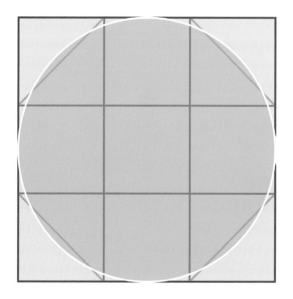

A square with inscribed circle and subdivided into parts by a grid showing the ancient method of approximation described in the Rhind Papyrus for calculating a the area of a circle.

Rhind/Ahmes Papyrus involves working out the area of a circle. They did this by first enclosing the circle to be measured inside a square and then drawing an octagon inside the square, conforming as closely as possible to the circle. By subtracting the areas of the triangles formed between the square and octagon they could calculate the area of the octagon, which was reckoned to be a close enough approximation to the area of the circle. This method actually yields an approximation of *pi*, the ratio of a circle's circumference to its diameter (more on this later!), of about 3.16, which is fairly close to the actual value of 3.14159...

THALES SHOWS THE WAY

The development of modern mathematics was influenced hugely by the Greeks, who were in turn influenced by the Egyptians. The first person we know of who approached mathematics in the sense we understand it today – following a set of abstract principles in a logical progression – was Thales of Miletus.

Thales (*c*.625–547BC) lived in what is now Turkey. He is often cited as one of the first to take a scientific approach to explaining the world, seeking to find answers for the workings of natural phenomena rather than seeing the actions of the gods behind them. Along with many other Greeks, Thales travelled to Egypt to study the practical skills of their geometers. A thousand years after Ahmes copied out his papyrus, Thales would have watched the *harpedonaptai* at work, measuring lengths and forming angles with their knotted ropes, and he brought his observations back to Greece.

Thales was the first to put forward mathematical theorems, which he 'proved' through observation and induction, showing that his theorems were correct by repeated experiments. These theorems were fairly basic (see box), but his work marked a radical new theoretical approach to mathematics, which before this had been almost purely practical, and led the way towards Pythagoras' development of maths as a science. Thales showed how geometry could be thought of in the abstract and yet its principles applied to the real world.

A good illustration of the leap forward in mathematical thinking made by the Greeks is the story of Thales and the pyramid. When Thales visited the Great Pyramid at Gizeh the structure was already 2,000 years old, but no one knew just how tall it was. Thales solved the problem using the principle of similar triangles.

Similar triangles have the same angles and proportions, but are not the same size. One version of the story had Thales putting a stick in the ground and recording the moment in the day when the length of the shadow cast by the stick was equal to the length of the stick. He reasoned that at that same moment the shadow cast by the pyramid would be equal to the height of the pyramid. The rope stretchers were able to tell him the width of the pyramid so all he had to do was add half that to the length of the shadow he could see, and he had his answer.

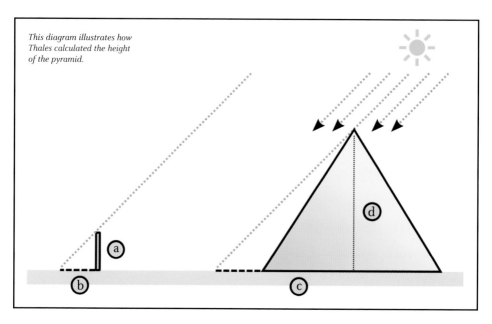

This diagram illustrates how Thales calculated the height of the pyramid.

Thales' theorem

Thales' theorem states that an angle in a semicircle will be a right angle. That is, if you take the diameter of a circle as the base of a triangle and then draw the other two sides of the triangle from any point on the perimeter of the circle, the angle opposite the base will always be a right angle.

If A to C is the diameter, then the angle at B will be a right angle.

ENTER EUCLID

Around 300BC, the principles of geometry as they were then known were set out in a 13-book work called *The Elements of Geometry*. Thomas Heath, who produced the standard English translation of *The Elements* in 1908, pronounced it a 'wonderful book' that would 'doubtless remain the greatest mathematical textbook of all time'. *The Elements* ranks second only to The Bible in the number of times it has been translated and reprinted.

The mathematician who produced this opus was said to be a Greek called Euclid of Alexandria. We can't know how much of what appears in *The Elements* is down to Euclid's original thinking and how much is a gathering together of the thoughts of others. A citation by Proclus, one of the last of the major Classical philosophers, in his book *Commentary on The Elements*, leads to the assumption that Euclid existed and was credited with having written *The Elements*. But, beyond that, we know little of the man himself apart from a few brief mentions by his contemporaries. It is possible that Euclid was simply the principal member of a team of scholars who compiled the book. Whoever it was who formulated the definitions, axioms, theorems and proofs of *The Elements*, they built the foundations of geometry for centuries to follow and may rightly be considered the 'Father of Geometry'.

The problems Euclid set out to tackle included plane geometry, solid geometry and number theory, including prime numbers. The importance of *The Elements* lies in the approach Euclid took to his problem solving. He began by setting out a number of postulates. These are mathematical laws and propositions that are assumed to be true and require no proof. From the building blocks of these five fundamental principles, Euclid set out to construct, step by step, his system of geometry.

Euclid's Postulates

1. A straight line segment can be drawn joining any two points.
2. Any straight line segment can be extended indefinitely in a straight line.
3. Given any straight line segment, a circle can be drawn having the segment as radius and one endpoint as centre.
4. All right angles are congruent.
5. If two lines are drawn which intersect a third in such a way that the sum of the inner angles on one side is less than two right angles, then the two lines inevitably must intersect each other on that side if extended far enough. This postulate is equivalent to what is known as the parallel postulate.

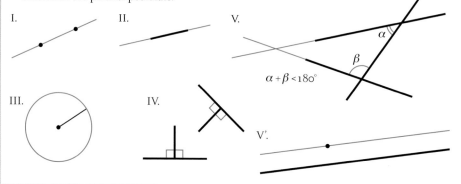

Euclid was one of the first mathematicians to use logic to prove his theories. The idea of requiring rigorous proof of a proposition is one of the fundamental principles of mathematics.

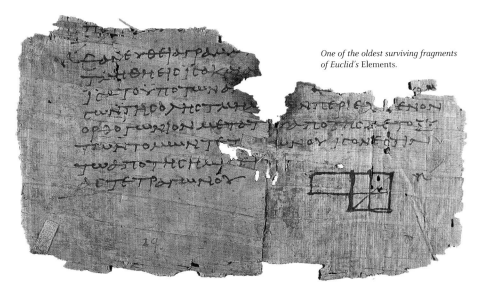

One of the oldest surviving fragments of Euclid's Elements.

TRIGONOMETRY

Trigonometry is the study of the relationships between the angles in a triangle and the lengths of its sides. Plane trigonometry, of special interest to surveyors and map-makers, involves triangles in which the angles all add up to 180°. Spherical trigonometry, of special interest to astronomers, involves triangles on the surface of a sphere, where the angles add up to more than 180°.

The Babylonians were measuring angles in degrees sometime before 300BC. They were the first to give co-ordinates for stars. They used the ecliptic, the Sun's apparent path through the sky over a year, as their base circle in the celestial sphere. They measured the longitude in degrees anticlockwise from the vernal point, which is the Sun's position at the vernal (or spring) equinox (when night and day are of equal length), as seen from the North Pole. And they measured the latitude in degrees north or south of the ecliptic.

Hipparchus (*c*.190BC – *c*.127BC) was a Greek astronomer and mathematician who made many important astronomical discoveries, including the 'precession of the equinoxes'. This is the change in the Sun's apparent position at the vernal equinox that occurs over long periods of time. It is caused by changes in the Earth's axis of rotation – in other words, the Earth's wobble as it spins. He noticed this by comparing the Sun's position in his day with its position on older charts. He also calculated the length of the year to within 6½ minutes. Perhaps his greatest achievement was to compile the first-known star catalogue. The Babylonians, Egyptians and earlier Greeks had all studied astronomy before him and determined the positions of many stars on the celestial sphere, but Hipparchus' catalogue, completed in 129BC, was an extraordinary work for its time. It listed around 850 stars, setting their positions in terms of celestial latitude and longitude with greater accuracy than anyone had before, and using a system of magnitudes to record their brightness that is similar to the one in use today.

It was his astronomical work that spurred Hipparchus to develop certain branches of mathematics. He produced an early form of trigonometry and set out a 'table of chords'. A 'chord' is the line joining two points on a unit circle that corresponds to the given angle at the centre. The chord of an angle *AOB*, where *O* is the centre of a circle and *A* and *B* are two points on the circle, is simply the straight line, *AB*. The length of the chord is proportional to the radius of the circle. The table would have helped him greatly in his astronomical calculations. According to Theon of Alexandria, who lived around 500 years later, Hipparchus wrote a 12-book work on chords, since lost. If so, that would make it the earliest known work of trigonometry.

The chord is closely related to the *sine* used in the modern version of trigonometry. A *sine* is half a chord or, put a different way, the *sine* of an angle is half the chord of twice the angle. Hipparchus is known to have developed a method of solving spherical triangles. (A spherical triangle is a figure formed on the surface of a sphere by three intersecting arcs.)

It is also generally agreed that the theorem in plane geometry known as 'Ptolemy's theorem' (which describes the relationship between the sides and diagonals of a cyclic quadrilateral – a four-sided figure inscribed in a circle) was originally due to Hipparchus and was later copied by Ptolemy (Claudius Ptolemaeus, *c.*AD100–178). This theorem states that the product of the diagonals (their lengths multiplied together) equals the sum of the products of the opposite sides.

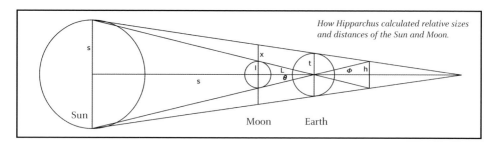

How Hipparchus calculated relative sizes and distances of the Sun and Moon.

Regiomontanus

The German scholar, Johann Müller of Königsberg, better known as Regiomontanus (1436–76), was perhaps the most capable mathematician of the 15th century. His main contribution to mathematics was in the area of trigonometry. It was largely through his efforts that trigonometry came to be considered an independent branch of mathematics. His book *De triangulis*, in which he described much of the basic trigonometry that is now taught in schools, was the first great book on the subject to appear in print.

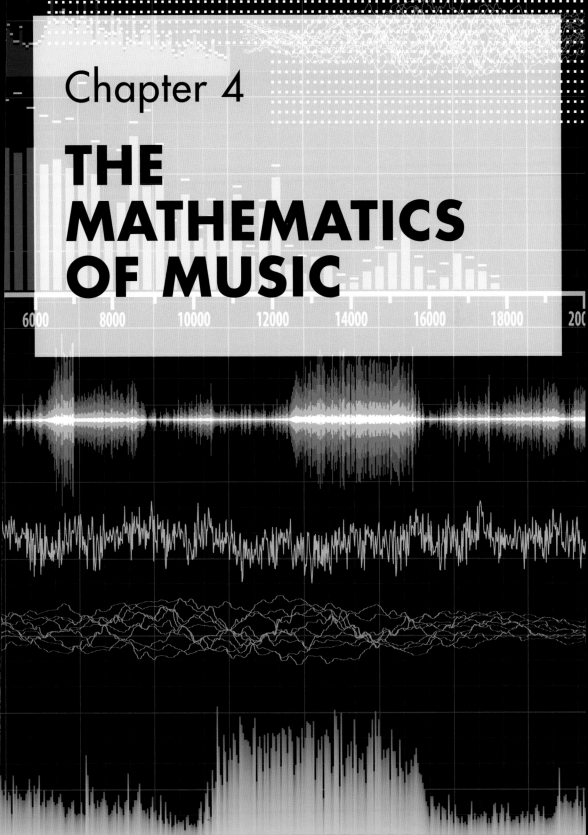

Chapter 4

THE MATHEMATICS OF MUSIC

6000   8000   10000   12000   14000   16000   18000   200

The mathematics of music

Mathematics and music might appear to be entirely different things. You're not likely to find yourself tapping your feet to an equation, unless perhaps you are a mathematician. Yet just think about how deeply numbers are involved in music. Rhythm, scales, intervals, time signatures, tone and pitch all have a mathematical element. How did mathematics and music become linked? And how did mathematics solve the problem of tuning?

'Musical form is close to mathematics – not perhaps to mathematics itself, but certainly to something like mathematical thinking and relationships.'

Igor Stravinsky, composer.

Music timeline	
*c.*6th century BC	Pythagoras carries out his experiments on harmony.
AD1581	Vincenzo Galilei proposes a system of equal intervals for tuning the lute.
1822	Joseph Fourier proves that any continuous function could be produced as an infinite sum of sine and cosine waves.

SOUNDS AND SIZES

According to legend, Pythagoras was passing a forge when his attention was captured by the sounds of the blacksmith's hammer. He noticed that a hammer weighing half as much as another produced a note an octave higher as it struck the metal. Although this event may never have taken place, it is indisputable that Pythagoras conducted experiments into the relationship between the size of an object and the tone it produced. These experiments included plucking strings of different lengths and striking vessels filled with varying quantities of liquid to see how the notes changed.

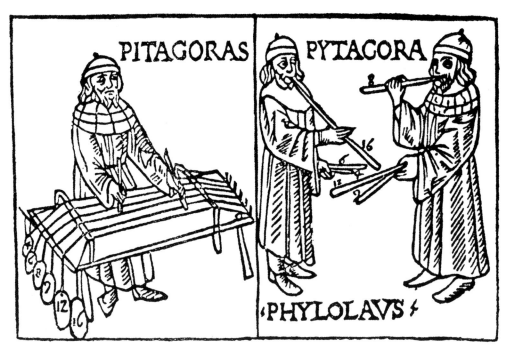

Woodcut of Pythagoras experimenting with sound.

As a result of his experimentation, Pythagoras established a mathematical relationship between object and sound. He discovered that the intervals between harmonious musical notes always have whole-number ratios. For example, if you take two taut strings of the same length, the same material and the same tension, they will both make the same sound. If one is twice the length of the other, the short string will vibrate with twice the frequency of the longer string and the resulting notes are an octave apart. The ratio of the octave, as Pythagoras discovered, is 2:1.

If one string is a third the length of the other the ratio is 3:2 and the interval, or difference in the notes produced, is a fifth.

A ratio of 4:3 (from a quarter-length string) produces an interval of a fourth. Sounding an octave, a fifth and a fourth together produces a harmonious sound that is pleasing to the ear – a 'musical chord'. Non-whole number ratios, on the other hand, tend to produce dissonant sounds.

Pythagoras had succeeded in explaining a natural phenomenon, in this case sound, in terms of numbers. This had never been done before. Pythagoras' discoveries led him to the belief that musical harmonies were reflected in the Universe as a whole, and that numbers and their relationships could explain all things. Pythagoras became convinced that the whole universe was based on numbers, and that the planets and stars moved according to mathematical equations, which corresponded to musical notes, producing the 'Music of the Spheres', an idea that persisted for over 2,000 years.

Sounds are produced by vibrations, the higher the frequency of the vibrations

Pythagoras celebrates sunrise *by Fyodor Bronnikov.*

(measured in Hertz, Hz), the higher the pitch of the sound we perceive. The musical note A above the middle C of a piano has a frequency of 440Hz (this is known as the reference tone – the one other instruments tune to); doubling the frequency to 880Hz produces an A that is an octave higher.

Prime problems

The frequency of a standard A in the early part of the 20th century was 439Hz. In May 1939, an international conference in London agreed the present standard of 440Hz. Why did this happen? The answer may have something to do with the arrival of radio broadcasting. Radio brought concert performances to a growing listening public. The BBC generated a tuning note using an oscillator controlled by a piezo-electric crystal that vibrates with a frequency of 1,000,000Hz. This was divided and multiplied in a number of steps to produce the required frequency of 440Hz – 439, being a prime number, could not be generated by this method.

PYTHAGOREAN TUNING

The oldest way of tuning the 12-note chromatic scale used by the Greeks is called Pythagorean tuning. It is built on a series of perfect fifth intervals using the 3:2 ratio. If the first interval step is 3:2, and the next is 3:2 above that, then the third note's ratio to the first note will be 9:4. But this means the pitch of the third note is more than an octave above the first note; to bring it within the same range we halve it, reducing it by an octave, and giving a ratio of 1.125:1 (or 9:8). So, we now have three notes, the base note, a note with a frequency that is 1.125 times the frequency of the base note, and one that is 1.5 times the frequency. We can carry on using this method to generate additional notes. Continuing for a total of 12 steps produces all 12 notes of the musical scale, finishing an octave higher than we started.

Unfortunately, this is a flawed system. The numbers simply don't add up. Repeatedly applying the ratio of 3:2 results in the 12th note actually being not quite an octave above the first note. Rather than being in a ratio of 2:1, it is in fact in a ratio of 2.027:1 with the base note. This didn't matter to the Greeks who simply avoided the slightly out-of-tune notes. The limitations of the Greek tuning system were exposed when music-making became more sophisticated. This led to the adoption of the equal temperament system. Unlike the Greek whole-number system, this was built on *irrational numbers* (see box, page 37).

A musician tuning a violin.

EQUAL TEMPERAMENT

The equal temperament system of tuning balances the musical scale so that 12 notes are equal to an octave. Effectively, what it does is to spread out the tuning errors inherent in the Greek system. It doesn't eliminate the errors but it does reduce them, so that although every note is still slightly out of tune it isn't unacceptably so.

Florentine music theorist Vincenzo Galilei (father of the famous astronomer Galileo) proposed a system of equal intervals for tuning the lute in 1581. French mathematician Marin Mersenne wrote of a similar tuning system in 1636. By the late 18th century, equal temperament tuning had been widely adopted by the musicians and instrument-makers of France and Germany and soon spread across the rest of Europe.

To achieve the 2:1 ratio of the octave in 12 steps requires the ratio between each step to be such that when multiplied by itself 12 times it results in a ratio of 2:1. In other words, $x^{12} = 2$, which means that equal temperament is built from multiples of the twelfth root of two (an irrational number, see box), a ratio of approximately 1.0595:1. In this

Irrational numbers

The Pythagoreans believed that all numbers could be described as the ratio of two whole numbers, or 'integers'. These are known as rational numbers, from the word 'ratio'. It must have come as something of a shock to discover that there were numbers that just couldn't be expressed as simple proportions.

Consider a right-angled triangle formed by cutting a square along its diagonal. Two sides of the triangle are 1 unit long, the third side, according to the theorem named after Pythagoras, must be equal to the square root of 2. So how long is it? Hippasus, a pupil of Pythagoras, tried to work it out but found that he just couldn't express it as a ratio between two integers. According to legend, Pythagoras was so horrified by this finding that he took poor Hippasus out in a boat and drowned him! It is somewhat ironic that a consequence of the theorem that bears Pythagoras' name resulted in such an undermining of his philosophy.

Obviously $\sqrt{2}$ can't be a whole number as it has to lie somewhere between 1 and 2 and is – to save you working it out – 1.4142135623730950..., the decimal expansion continuing on into infinity. There is no whole number fraction that we can find that equates to this number. As there is no equivalent ratio it is, therefore, irrational.

A cartoon of Hippasus drowning while contemplating irrational numbers.

tuning, the ratio of a fifth is 1.498:1, which, to the human ear is, purely by coincidence, indistinguishably close to the Pythagorean 1:5 fifth. A third, however, is 1.26:1, which is audibly different from the Pythagorean ratio of 4:3 (1.25:1). The sophistication of the music we enjoy today is thanks to a number that the great Pythagoras could scarcely bring himself to believe existed.

SOUND AND FOURIER

A steady, pure tone of a single pitch is produced by a sine wave, a smooth repetitive oscillation. The sounds produced by musical instruments are much more complex. The quality, or timbre, of the sound of a musical instrument results principally from the harmonic content, the number and relative intensities of the upper harmonics present in the sound. The result is a sophisticated waveform with a variety of waves interfering with each other.

French scientist Jean-Baptiste Joseph Fourier (1768–1830), while researching the way heat transfers from one place to another, showed that, no matter how complex the waveform, it could be broken down into its constituent sine waves, a process called Fourier analysis. A sound wave can therefore be characterized in terms of the amplitudes of the constituent sine waves it is composed of. This set of numbers is sometimes referred to as the harmonic spectrum of the sound. Once you know the harmonic content it is possible to reverse the process and build up a synthesized version of the original sound using tone generators to create the constituent sine waves.

A variety of modern technologies such as radio communication, noise-cancelling headphones and speech recognition software rely on Fourier analysis.

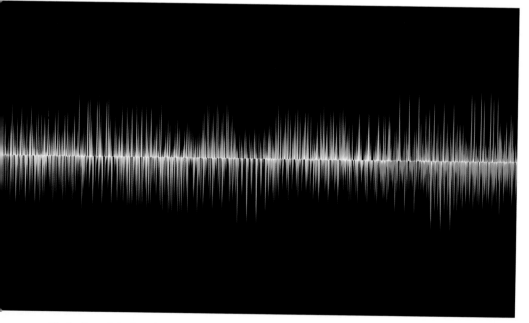

The Fourier analysis allows the breakdown of complex waves into simpler components.

Chapter 5

GOING AROUND IN CIRCLES – THE PATH TO *PI*

Going around in circles – the path to *pi*

In terms of geometry, all circles are similar. In other words, no matter the size of the circle, the ratio of its diameter to its circumference is always the same. This ratio, called *pi*, and denoted by the symbol π, is the most famous of all mathematical constants.

The first theoretical calculation of this ratio seems to have been carried out by the Greek mathematician Archimedes of Syracuse. Since then mathematicians have employed a great deal of problem-solving ingenuity in trying to establish more and more accurate values for *pi*. *Pi* is an irrational number with no 'exact' value. It is an infinitely long string of digits, stretching out as far as our mathematical ability can go, and beyond.

Although it is a quest with no end in prospect, the efforts to achieve a greater mathematical understanding of *pi* over the centuries have led to many scientific and technological advances.

Calculations using Pi to determine the area of a circle.

A *pi* timeline

*C.*2,000BC	The Babylonians and Egyptians determine *pi* to be just over 3.
*C.*250BC	Archimedes gives an approximation of *pi* as $^{22}/_{7}$ and calculates it to a value of 3.1418
AD1706	Welsh mathematician William Jones is the first to use the symbol π. The notion was popularized by Swiss mathematician Leonhard Euler in the mid-18th century.
1768	Johann Lambert proves that *pi* is an irrational number and that its decimal expansion carries on into infinity without repeating or following any predictable pattern.
1882	German mathematician Ferdinand von Lindemann (1852–1939) proves that π isn't just irrational, it is also transcendental. A transcendental number is one that can't be expressed in any finite series of arithmetical or algebraic operations.

Prehistoric circles

Circles have fascinated people of all cultures across the entirety of human history. Prehistoric rock art from around the world commonly features circular markings. Stonehenge and other megalithic monuments are set out in a circular pattern. It is generally believed that Stonehenge had some astronomical purpose, as it aligns with important events such as the rising of the Sun at the winter and summer solstices. Anthony Johnson, an industrial archaeologist at Oxford University, believes it also provides evidence of sophisticated skills in solving problems in geometry.

Stonehenge in England.

The most complex geometrical feature of Stonehenge is an 87m (285 ft) diameter circle of pits marking the points of a 56-sided polygon, which has been created just inside the monument's perimeter earthwork. Using computer analysis, Johnson demonstrated that the polygon was created using nothing more than a rope and a post. He believes that the Stonehenge surveyors began by using a rope to create a circle, then laid out the four corners of a square on its circumference, before laying out a second similar square to create an inner octagon. The points of the octagon were then used as anchor points for a rope that was used to describe a series of arcs intersecting the circumference of the circle and thus eventually forming a huge 56-sided polygon. Johnson also showed that a 56-sided polygon is the most complex that can easily be created using this technique. He said that 'the builders of Stonehenge had a sophisticated ... knowledge of Pythagorean geometry 2,000 years before Pythagoras.'

The observation that the ratio of a circle's circumference to its diameter is the same for all circles, whatever their size, has been known for thousands of years. This ratio eventually became known as *pi* – symbolized by the Greek letter π. As we saw earlier (page 25), the Egyptians calculated its value at around 3.16. The Babylonians were also aware of the ratio, using a value of 3.12, which they calculated by inscribing a hexagon

Sarsen
Fallen sarsen
Bluestone (generic)
Fallen bluestone
Sandstone

An aerial plan of the Stonehenge site.

inside a circle and assuming that the ratio of the hexagon's perimeter to the circle's circumference was $^{24}/_{25}$.

Around 250BC, Archimedes calculated the area inside a curve by subdividing the area into a number of tiny strips. He calculated the area of each strip and added them all together to find the solution. This method was a direct precursor of integral calculus, developed by Isaac Newton and Gottfried Leibnitz nearly 2,000 years later (see pages 112–14).

Archimedes

Archimedes of Syracuse (*c.*287–212BC) is widely regarded as having been the ancient world's greatest scientist. His abilities as an engineer, physicist and mathematician were unparalleled and he made a lasting impact on the history of mathematics. As an engineer, he created the screw pump for irrigation that takes his name, along with a variety of applications of levers and pulleys. Perhaps he is best known for the Archimedes' principle, or the principle of buoyancy. The image of Archimedes jumping naked from his bath and shouting '*Eureka!*' ('I've found it!') is, true or not, enshrined in science history. In addition to his calculations of *pi*, his mathematics included finding the perimeter, area, and volume of many geometric shapes, such as spheres and cylinders.

Archimedes set out to establish the value of *pi* as accurately as he could. To calculate the value, he used a similar approach to that described in the Ahmes papyrus. However, instead of using an octagon, he used a 96-sided polygon. Archimedes approximated the area of a circle by calculating the areas of two polygons: one inscribed within the circle one outside it. The actual area of the circle lay somewhere between the areas of the two polygons, which gave upper and lower limits for the area of the circle. Archimedes knew that his method didn't yield an absolute value for *pi* but an approximation of it, which he calculated as 3.1418.

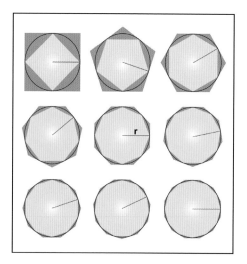

How Archimedes calculated the area of a circle.

Statue of Archimedes in Syracuse.

Archimedes' method of calculating *pi* was used by mathematicians across the world for the next 18 centuries, and especially in China, India and in the Islamic countries. In 1596, Dutch mathematician Ludolph van Ceulen took this approach just about as far as it could go when he calculated *pi* to 35 decimal places by using a polygon with approximately 4.6 billion billion sides! As he spent most of his life engaged in this heroic calculation it seems only right that the 35 places of *pi* were engraved on Van Ceulen's tombstone.

Ludolph van Ceulen's tombstone.

INFINITE SERIES

The development of infinite series techniques in the 16th and 17th centuries greatly enhanced the ability of mathematicians to approximate *pi* more efficiently. An infinite series is the sum (or much less commonly, product) of the terms of an infinite sequence, such as $\frac{1}{2}$, $\frac{1}{4}$, $\frac{1}{8}$, $\frac{1}{16}$ ... $\frac{1}{(2^n)}$. The first written description of an infinite series that could be used to compute *pi* was laid out in Sanskrit verse by the Indian astronomer Nilakantha Somayaji around AD1500.

Buffon's needle

The 18th-century French mathematician George Buffon suggested a curious method of determining *pi* that involved randomly dropping a needle on to a uniform grid of parallel lines. So long as the length of the needle, *l*, is less than the distance between the lines, the probability that the needle will fall across a line is $2l/\pi$. In 1901, Italian mathematician Mario Lazzarini tried out the experiment making 34,080 needle tosses and getting a result for *pi* of $^{355}/_{113} = 3.1415929$, which seems remarkably accurate – indeed some mathematicians have commented that it is suspiciously so. The odd choice of the number of tosses may be what gives the game away, suggesting that Lazzarini waited until his result was looking good before calling a halt.

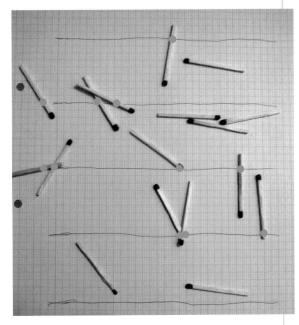

'Buffon's needle' method of calculating pi *demonstrated using matchsticks.*

In 1665, Isaac Newton used infinite series and the calculus that he and Leibniz had separately developed to compute *pi* to 15 digits. The first 100-digit approximation had been achieved by the early 18th century. A 620-digit approximation made in 1946 was the best attained without the aid of a calculator or computer. Using the computing power available today, *pi*'s value can be calculated into the trillions of digits. In November 2016, after 105 continuous days of computing, mathematician Peter Trueb calculated *pi* to a staggering 22,459,157,718,361 fully verified digits. To get an idea of just how lengthy this number is, if it was printed out it would fill several million 1,000-page books. The computer Trueb used was equipped with 24 hard drives, each containing six terabytes of memory, to store the vast quantity of data being produced.

How far do we need to go?

Engineers working to a high degree of tolerance find it invaluable to have a close approximation of *pi* to work with. But how close does it really have to be? Canadian mathematicians Jonathan and Peter Borwein calculated that a value of *pi* to 35 decimal places would be enough to determine the circumference of a circle large enough to encompass the entire visible universe to an accuracy of less than the radius of a hydrogen atom. In working out the course calculations for its interplanetary space probes, NASA uses a 15-place value of $\pi \approx 3.141592653589793$.

New Horizons spacecraft and Pluto.

Chapter 6

ZERO – THE NOTHING THAT WAS SOMETHING

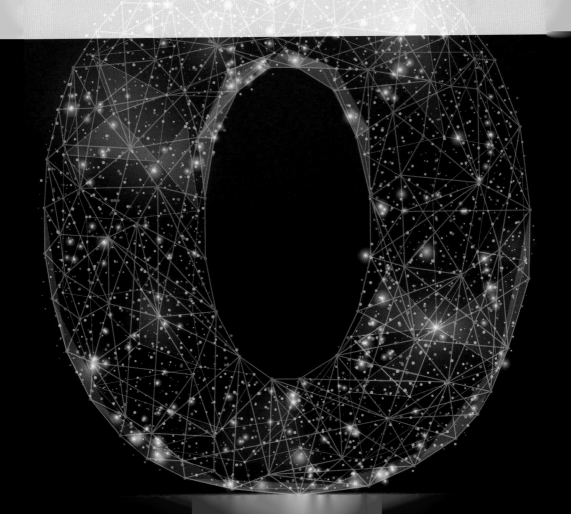

Zero – the nothing that was something

Numbers are used for counting things – that's obvious. But how do we count something that isn't there? How many oranges are there in an empty box? The conceptual leap to use zero as a number in its own right, rather than simply as a placeholder, is usually attributed to 5th-century Indian mathematicians, though it may have been in use for centuries prior to that.

(Above) *It is hard for us to imagine life without zero.*

The earliest recorded usage of a circle character for the number zero may be a 9th-century engraving in a temple in Gwalior in central India. The use of zero as a number revolutionized mathematics. It was, for example, essential to the development of René Descartes' Cartesian co-ordinate system and to Newton and Liebniz's developments of calculus.

The entrance to the temple of Gwalior.

Zero timeline

*c.*700BC	The Babylonians use zero as a placeholder in their number system.
AD628	Indian mathematician Brahmagupta uses zero and sets out rules for its use with other numbers.
*c.*600–680	Bhaskara, another Indian mathematician, uses 0 as an algebraic symbol and considers what division by zero means.
9TH CENTURY AD	The earliest known example of the zero symbol is engraved on a stone tablet in Gwalior, India.

THE NEED FOR ZERO

There is nothing intuitively obvious about a need for zero. When we first learn about numbers in school we start with 1, 2, 3, 4... Numbers are used for counting things – that's why humans invented them. Far from the purely mathematical abstract ideas we use them for today, numbers stood for something in the real world – five sheep, ten trees, two children. A number was simply a word that referred to a collection of objects. It was quite some time before anyone came up with the notion that we needed a number to express the idea of nothing being there.

Making use of nothing

There are two different, but equally important, uses for zero. One is as an empty place indicator in a place-value number system. 2001 is very different from 21 – we need the two zeroes in there to tell us that the 2 equals 2,000 and not 20. The second use of zero is as a number in itself.

You might imagine that inventing a place-value number system would, of necessity, mean adopting an empty place indicator too, but, as we have seen, the Babylonians appeared to manage very well without one for centuries. It wasn't until around 400BC that the Babylonians put two wedge symbols into the place where we would put zero to indicate an empty place. Other symbols were also used: an earlier tablet from the

An example of Babylonian cuneiform script. In the middle, four lines from the bottom, the double wedge symbol appears as a placeholder for the value of zero.

Mesopotamian city of Kush, thought to date from around 700BC, uses three hook symbols to mark an empty space in the positional notation, for example.

Whatever the symbol used, it only occurred between digits, never at the end of the number. It isn't really the use of zero as a number, it's more of a punctuation mark. We can only assume that context was enough to know whether, say, 30 was meant rather than 3. If that seems odd, consider that if you buy a cup of coffee and a cake and are asked for 'four thirty' you know its four pounds and thirty pence that's wanted – not four hundred and thirty pounds.

NOTHING TO THE GREEKS

Unlike the Mesopotamians, the Greeks did not adopt a place-value number system. The Greeks, as we have seen, were accomplished mathematicians, but their approach to the subject was entirely different. Greek mathematics was fundamentally about geometry. They used two number systems, both of them based on letters in the Greek alphabet, rather than using dedicated number symbols. This didn't allow for calculations to be made using the numbers themselves and it is thought that the Greeks had to use counters of some sort to work out sums. They were much more interested in things like line lengths and proportions.

There was, however, an important exception to this – the astronomers. When recording astronomical data, Greek mathematicians used the symbol 0 – the first use of the symbol we recognize today for zero. Some maths historians think it stands for omicron, the first letter of the Greek word for nothing. Others disagree, pointing out that omicron already represented 70 in their alphabet-based number system. Another idea is that it stands for 'obol', a coin of little value, used as a counter on a sand board. When the counter was removed, leaving a column empty, it left a mark in the sand that looked like 0. Whatever the explanation, the use of a placeholder zero didn't become established and eventually, perhaps fittingly, disappeared. Zero didn't reappear for a few centuries, this time in India.

THE MATHEMATICIANS OF INDIA

In 1881, an ancient manuscript was uncovered in the village of Bakhshali, in what is now Pakistan. Its 70 birch bark leaves have proved difficult to date but the best guess is that it was written around AD400. Various mathematical rules are laid out on the pages and it may represent the first documented use of the Hindu–Arabic numerals that evolved into the familiar number symbols we use today. It may also be the first fully realized decimal place-value number system, complete with a symbol for zero, in this case a dot. It even includes the idea of negative numbers, in the context of profit and loss.

BRAHMAGUPTA

The *Brāhmasphutasiddhānta* (Correctly Established Doctrine of Brahma), written in AD628 by the Indian mathematician Brahmagupta (AD598–c.665?) is one of mathematics' most important texts. It marked the first time that zero was elevated from place-marker to the status of a number in its own right.

Brahmagupta was an astronomer of note as well as a gifted mathematician. The *Brāhmasphutasiddhānta* featured important insights into algebra, number theory and geometry but perhaps most importantly it laid out a new set of rules for arithmetic that Brahmagupta had formulated to include the number zero.

The first zero

The first confirmed record of the use of the zero symbol we recognize today was written in AD876. It was found in an inscription on a stone tablet (part of the inscription is the date, which translates as 876) concerning the town of Gwalior. It refers to a garden that would produce enough flowers to allow 50 garlands per day to be given to the local temple. The number '50' appears almost as it does today, although the 0 is smaller and slightly raised.

Brahmagupta explained, for example, that if you subtract a number from itself you obtain zero. He also offered the following rules for addition that involve zero: the sum of zero and a negative number is negative, the sum of zero and a positive number is positive, the sum of zero and zero is zero. Similarly for subtraction:

- a negative number subtracted from zero is positive,
- a positive number subtracted from zero is negative,
- zero subtracted from a negative number is negative,
- zero subtracted from a positive number is positive,
- zero subtracted from zero is zero,
- a number multiplied by zero is zero.

All of which may seem rather obvious to 21st-century mathematical sophisticates, but remember that no one had really thought of this before.

INTO THE NEGATIVE ZONE

Brahmagupta didn't stop at zero. He continued into the hitherto virtually uncharted territory of negative numbers. If there was something counterintuitive about the need for a zero, how much more so about the need for a negative number? For example, how could a box contain minus three bananas?

As in the earlier Bakhshali manuscript, Brahmagupta couched his argument in terms of finance, using negative numbers to signify debt. Having minus three bananas meant I owed someone three bananas. Zero was the point at which the books balanced, with nothing owed on either side. What Brahmagupta had accomplished, perhaps for the first time, was to unite positive, negative and zero into a complete, coherent number system.

Among the many laws of arithmetic that he formulated, Brahmagupta grasped a concept that still eludes many school pupils today, namely that if you multiply two negative numbers together the result is a positive number. But if you multiply a positive number by a negative number the result will be negative.

Negative numbers. Brahmagupta was the first to include positive and negative numbers in a coherent number system.

THE DIVISION-BY-ZERO CONUNDRUM

Brahmagupta had little to say on the subject of dividing a number by zero and seemed rather baffled by the idea. It's not really surprising. What does dividing by zero actually mean? If I have 12 bananas and want to divide them into piles each containing zero bananas how many piles do I make? In 830, Mahavira attempted to update Brahmagupta's book. He wrote that a number remains unchanged when divided by zero, which is surely not right.

Three hundred years later, Bhaskara was still struggling to come to terms with division by zero. He decided that: 'A quantity divided by zero becomes a fraction, the denominator of which is zero. This fraction is termed an infinite quantity.' This is, of course, wrong too. If it was right, it would mean that zero multiplied by infinity would equal every number there is, which is absurd. The mathematicians of India seemingly couldn't bring themselves to accept that division by zero was meaningless or, as mathematicians have it, 'indeterminate'.

ZERO THE HERO

Early mathematicians may have been reluctant to embrace zero but it is hard to imagine life without it now. Modern science and mathematics simply could not do without zero. We have zero degrees on temperature scales; it separates the positive numbers from the negative ones; it's the point at which the axes on a graph cross over; it's a placeholder that lets us use numbers that are astronomically huge and infinitesimally small; in the binary system that underpins our computers, it's the 'off' position. It's even used in everyday language as we 'zero in on something', wait for 'zero hour' and show 'zero tolerance'. That nothing really is quite something!

Chapter 7

ALGEBRA – SOLVING THE UNKNOWN

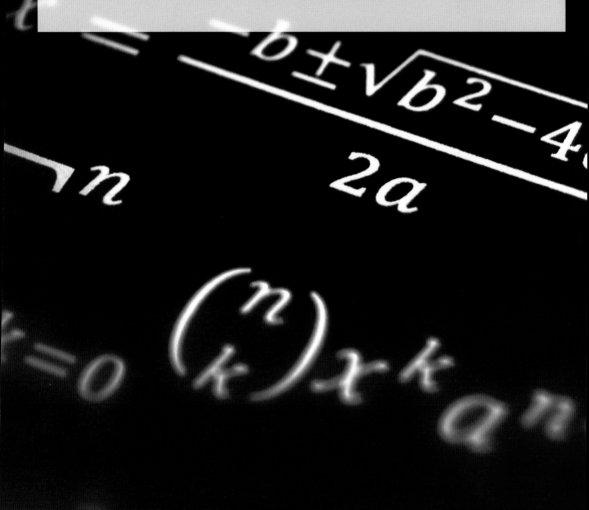

Algebra – solving the unknown

Algebra is the science of solving equations (basically, mathematical puzzles) when working with unknown quantities. Just like a chemical equation, a mathematical equation has to be in balance – one side must equal the other – so if you know what's on one side then, it stands to reason, you can work out what's on the other side. Algebra has many practical applications today, including in computing, finance and science.

Although the origins of algebra can be traced back to the mathematicians of ancient Egypt and Babylon, it was a book by the medieval Arab mathematician Abu Abdullah Muhammad ibn Musa al-Khwārizmī who brought it to its first flowering. It would be developed further by the mathematicians of the Renaissance, who discovered how to solve cubic equations, and by Descartes' discovery of a way to link algebra to geometry.

Algebra timeline

*C.*1950BC	The Babylonians work out how to solve quadratic equations.
AD250	Diophantus of Alexandria publishes *Arithmetica*.
*c.*830	Al-Khwārizmī publishes his magnum opus on '*al-ğabr*' (algebra).
1535	Niccolò Fontana works out a solution for cubic equations.
1572	Rafael Bombelli introduces the idea of imaginary numbers.
1591	François Viète uses letters to denote known and unknown quantities.

EARLY EQUATIONS

Although it wasn't always known as algebra, the problem of solving equations has been around for a long time. The earliest surviving mathematical text from Egypt, the Rhind Papyrus (see page 22), provides ample evidence of the ability of the Egyptians to solve simple equations with unknown quantities of the $4x + 3x = 21$ variety. In more practical terms, Egyptian mathematicians could calculate the length and width of a field provided they already knew its perimeter and area. However, problems were stated and solved verbally – symbols weren't used.

Surviving clay tablets show that the Babylonians were able to solve equations in which the unknown quantities included squares (quadratic equations) and cubes (cubic equations). Problems such as these were accompanied by the procedure that needed to be followed in order to solve them. There was no attempt made to establish a general set of rules that would allow similar problems to be solved. Like those of the Egyptians, many of the problems dealt with practical situations, such as the dividing up of land, rather than being abstract puzzles.

A Babylonian clay tablet depicting mathematical problem-solving, including the use of quadratic equations, for determining the area of a field.

The Greeks, with their geometry-focused approach to mathematics, had very little part to play in the history of algebra and equation-solving. They did not use symbols in their calculations and the idea of equations was quite alien to their concept of mathematics.

The 3rd-century-AD Greek mathematician Diophantus of Alexandria was an early innovator in what would come to be called algebra, developing original methods for solving problems that today we would describe as linear or quadratic equations. But he still adhered to the Greek concept of mathematics, which had no place for negative numbers or zero. Any problem with a negative-number solution was simply absurd as far as Diophantus was concerned. Like the Egyptians and Babylonians before him, Diophantus offered *ad hoc* solutions to specific problems but made no attempt to establish generally applicable techniques for problem-solving. He never went beyond the first solution he found, even though many of the problems he tackled had many, sometimes even infinite, solutions.

One thing Diophantus did do was to introduce a sort of symbolism, but it was more of a convenient shorthand than a flexible set of symbols that could be reordered within the equation as a problem-solving mathematician might do today. His *Arithmetica*, a collection of problems giving solutions to equations, was the most obviously algebraic work in all Greek mathematics.

The Golden Age of Islam, the period from the mid-7th century to

DIOPHANTI
ALEXANDRINI
ARITHMETICORVM
LIBRI SEX,
ET DE NVMERIS MVLTANGVLIS
LIBER VNVS.

CVM COMMENTARIIS C. G. BACHETI V. C. & obseruationibus D. P. de FERMAT *Senatoris Tolosani.*

Accessit Doctrinæ Analyticæ inuentum nouum, collectum ex varijs ciusdem D. de FERMAT Epistolis.

TOLOSÆ,
Excudebat BERNARDVS BOSC, è Regione Collegij Societatis Iesu.
M. DC. LXX.

The title page of Diophantus' Arithmetica *from 1670.*

the mid-13th century, saw the spread of Greek and Indian mathematics to the Muslim world, as ancient Greek and Hindu works were translated into Arabic. This led to the development of an Islamic culture of mathematics and astronomy that brought together methods and ideas from India and elsewhere and added further improvements and innovations devised by Arabic scholars.

One of the prime innovators was Al-Khwārizmī, a scholar of the legendary *Bait al-Hikma*, or 'House of Wisdom', in Baghdad, where he worked as an astronomer, geographer and mathematician. Around AD830, he published *Al-kitāb al-mukhtaṣar fī ḥisāb al-ǧabr wa'l-muqābala* (The Compendious Book on Calculation by Completion and Balancing), and in the process gave us the phrase *al-ǧabr*, which became latinized as 'algebra'. He also adopted the practice, probably learned from Hindu texts, of using a small circle as placeholder rather than leaving a blank space in a column of numbers. The Arabs called this circle *sifr*, or 'empty', from which we get both 'cypher' and 'zero'.

Artist's impression of Bait al-Hikma.

Al-Khwārizmī wrote his book with the stated intent of teaching 'what is easiest and most useful in arithmetic'. In common with the standard practice of the time, Al-Khwārizmī's approach to algebra was rhetorical, that is to say, it was entirely in prose without any symbolic notation. What later mathematicians would label *x* to signify an unknown quantity, Al-Khwārizmī called *shay*, meaning 'the thing'.

Where he broke new ground was in approaching the subject in terms of abstract principles, thus parting company with the likes of the Egyptians and Diophantus, who only offered solutions tailored to particular problems. Al-Khwārizmī provided what was probably the first serious analysis of equations, opening them up for the scientists and bureaucrats of the time to use as tools to solve their practical and financial problems.

What's in a name?

Al-Khwārizmī's book on algebra was translated into Latin and, along with another he wrote on the Indian decimal system, was so widely disseminated that his name became part of the language of science and mathematics. Al-Khwārizmī became 'Alchoarismi' then 'Algorismi' and eventually 'algorithm' – a set of rules to be followed in calculations or other problem-solving operations, something that Al-Khwārizmī was also adept in developing. More than a thousand years after Al-Khwārizmī, every computer program would be encoding algorithms.

RESTORATION AND REDUCTION

Barbershops in medieval Spain advertised themselves as *Algebrista y Sangrador*. It didn't mean they'd tackle your maths problems for you, it meant 'Bonesetter and Bloodletter' – which used to feature in a barber's skillset along with shaving and haircutting. The root of *algebrista* is the Arabic *al-ğabr*, which means restoration, or reunion. In *hisāb al-ğabr wa'l-muqābala*, Al-Khwārizmī described restoration as the process by which $x = y - z$ becomes $x + z = y$, in other words turning a negative term into a positive one by resetting it on the other side of the equals sign. Reduction is the process of turning the equation $x = y + z$ into $x - z = y$. Both are examples of the general rule that whatever happens on one side of the equation must also happen on the other side. In the first equation, z was added to both sides; in the second equation, it was subtracted from both sides.

Postage stamp depicting Al-Khwārizmī.

X the unknown

French mathematician François Viète (1540–1603) was the first, in 1591, to introduce the use of letters to denote both the coefficients and the unknowns in an equation. This marked the beginning of a new type of algebra, one in which ideas could be expressed in terms of abstract formulas and general rules.

The standard algebraic notation we use today was introduced by René Descartes in his *La géométrie* (see page 103). He used letters from the beginning of the alphabet, *a, b, c* and so on, for known quantities, and *x, y, z* for unknown quantities. Apparently, when *La géométrie* was being typeset, the printer found himself short of letters and asked Descartes if it made any difference if he used *x, y* or *z*. Descartes replied that any of the three would do and the printer decided to use *x*, as it was less frequently used elsewhere. It was the decision of that unknown printer that resulted in us having X-rays, X-Files and the X Factor.

The variable x

As well as signifying an unknown quantity, x, or any other place-holding symbol, can also be used to signify a quantity that can have changing values. Using algebraic equations is immensely useful for scientists wishing to formulate general laws for the way things behave. For example, in Newton's force law $F = ma$: force (F) equals mass (m) times acceleration (a). We know that the three quantities are always going to have the same relationship to each other so in any specific case, if we know two values we can calculate the third.

Al-Khwārizmī found a way of solving quadratic equations that is still taught in high schools today. In case you've forgotten, or weren't paying attention, it goes like this. You begin with an equation of the form:

$ax^2 + bx + c = 0$

where a, b and c are any numbers

and the solutions are obtained using the equation

$$x = \frac{-b + \sqrt{b^2 - 4ac}}{2a}$$

Any quadratic equation can be solved by applying Al-Khwārizmī's formula. But other types of equation are not so easily solved. Cubic equations, which include the term x^3, have three different solutions, and quartic equations, with x^4 and four possible solutions, are even more of a problem. Mathematicians, therefore, continued to look for more equation-solving techniques. In 1070, Omar Khayyam, perhaps better known today as a poet through the works known as his *Rubáiyát*, made major steps forward when he wrote a *Treatise on Demonstration of Problems of Algebra*, classifying cubic equations that could be solved by conic sections. But there was still plenty of problem-solving to be done.

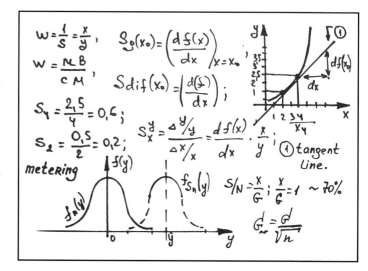

An example showing algebraic notations.

CUBIC COMBAT

The mathematicians of the Renaissance period were highly competitive, challenging each other to public contests in which they would set each other perplexing problems to solve. Reputations could be made or destroyed, fortunes won or lost. The up side of this gladiatorial maths was that real progress was made towards solving cubic equations.

With so much riding on these contests, new techniques were guarded jealously. Scipione del Ferro (1465–1526) was rumoured to have worked out how to solve cubic equations, but in the febrile atmosphere of competitiveness at the time this knowledge could provide a valuable edge against rival mathematicians. So Scipione kept his method a closely guarded secret. Before he died he entrusted it to just a handful of people, including his assistant, Antonio Maria Fior, and even he only had the solution to one of the three types of cubic equation.

NICCOLÒ THE STAMMERER

Niccolò Fontana (1499–57) was a self-taught mathematician from Brescia, Italy. He was known as 'Tartaglia' (the Stammerer) after a sabre cut to the mouth at the age of twelve left him with a speech impediment. Among his important contributions to mathematics was the translation into Italian of Euclid from the original Greek, correcting errors that had crept in when it had been translated from the Arabic version. Where he really forged his reputation, however, was in his skills in competition. In 1535, he took on Scipione's assistant, Fior, in a contest centred on the cubic equations.

The contest involved each

Niccolò Fontana 'Tartaglia'.

participant setting questions for the other, with forty to fifty days allowed to present their solutions. Fior set problems for the 'unknowns and cubes equal to numbers' form of cubic equation. It was the solution to this that had been gifted to him by Scipione and he was confident that he alone knew the secret.

As it happened that was true when he'd set the problems but, with just a week before solutions had to be presented, inspiration struck Fontana. Not only did he arrive at a solution to Fior's problems, he also arrived at a general solution for all types of cubic equation. It was an astonishing achievement. Up until then, most mathematicians had declared solving cubics to be an impossible task. In keeping with the paranoia of the times, Tartaglia encoded his solution in the form of a poem in an attempt to make it more difficult for other mathematicians to steal it.

THE DOUBLE-DEALING CARDANO

Unsurprisingly, Fontana's breakthrough attracted attention. One man who took an interest was Gerolamo Cardano (1501–76). A keen gambler (see page 80) as well as a first-rate mathematician and physician, Cardano attempted to wheedle the secret of the cubics out of Fontana. At first rebuffed, he eventually succeeded by promising the impoverished Fontana an introduction to a wealthy patron.

Working with his brilliant student Lodovico Ferrari (1522–65), Cardano set to work on expanding Fontana's solution. Ferrari realized that he could use a similar method to solve quartic equations (equations with terms including x^4), too. Within the confines of their geometry-based mathematics, the Greeks would have found quartics inconceivable. If squares represented areas, and cubics volumes, quartics were four-dimensional exotica.

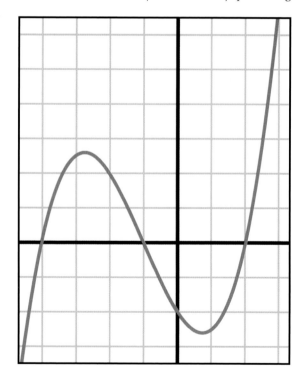

Cubic equations generally have a graph of this shape.

Having promised Fontana that he would not publish his work, Cardano happened to see del Ferro's unpublished solution for the cubic equations, which was dated before Fontana's. He argued that this meant that his promise to Fontana could legitimately be broken, and he included Fontana's solution to the cubics in his book of 1545, *Ars magna*, along with Ferrari's solution of the quartics. Although both Fontana and Ferrari were acknowledged in Cardano's book, Fontana was furious and began a battle with Cardano over the publication.

Ferrari's understanding of cubic and quartic equations outstripped Fontana's. When another mathematical duel took place between the two, Fontana soon realized that he was outmatched and withdrew from the competition. His reputation could not survive the ignominy and he became effectively unemployable, eventually dying impoverished and unknown. Adding posthumous injury, even today, the solution to cubic equations is usually known as Cardano's Formula and not Fontana's.

IMAGINARY NUMBERS

One of the problems that Cardano came across while working with cubics was that the formula sometimes involved square roots of negative numbers. This seemed an impossibility: after all a negative number multiplied by itself produces a positive result, so how could there be a square root of a negative number? In other words, the equation $x^2 = -1$ had no solutions.

Nonetheless, Cardano and other algebraic researchers found expressions such as $\sqrt{-1}$ appearing with increasing frequency as they worked through solutions for cubics and quartics. Had they gone wrong somewhere? Sometimes they found that if they persisted with the calculation they came across a cancelling expression such as $\sqrt{-1} \times \sqrt{-1}$, which could be replaced by -1. Puzzlingly, the solutions that came via this path through impossibility were still correct.

The problem was solved in 1572 when Rafael Bombelli published his book *L'Algebra*. In it, he set out the rules for an extended number system that included numbers such as $\sqrt{-1}$. René Descartes would later dismiss such numbers as 'imaginary' and the term stuck. Still without actually recognizing them as genuine numbers, the mathematicians of the day began working with imaginary numbers with more and more confidence, especially as it was recognized that Bombelli's rules were sound.

THE IMAGINARY *i*

In the 18th century, the great mathematician Leonard Euler (see pages 128–9) gave √−1 the name '*i*' or the 'imaginary unit', by which it is still known today. Other imaginary numbers are multiples of *i*. Bombelli's number system included yet another type of number. Alongside the real numbers (such as 5, −3 and π) and the imaginary numbers, there were also the complex numbers, formed by combining the first two, such as *a* + *bi*, where *a* and *b* are any real numbers, and $i = \sqrt{-1}$. Ultimately, as they probed further into the realm of the complex numbers, mathematicians began to realize just how powerful a problem-solving tool they had uncovered.

COMPLEX GEOMETRIES

The complex numbers led to a revolution in mathematics. In 1797, Carl Friedrich Gauss (1777–1855) (see also page 131) announced his proof that any equation built from real numbers could be solved using complex numbers. There were gaps in Gauss' theory, however. What of equations built from complex numbers? Would their solution require the number system to be extended yet further?

In 1806, Robert Argand solved the problem in a particularly elegant and insightful way. Any complex number, *z*, can be written, as stated, in the form *a* + *bi*, where a is the real part of *z* and *bi* is the imaginary part. Argand realized that the number system can be represented geometrically. If we consider (*a*, *b*) as Cartesian co-ordinates we can begin to explore the geometry of complex numbers. If the real numbers are drawn along the *x*-axis and the imaginary numbers along the *y*-axis then the whole plane between them becomes the realm of the complex numbers. This depiction of complex numbers is called the Argand diagram. In this representation *i* is interpreted as a rotation of the plane through 90°.

Argand proved that the solution for every equation built from complex numbers could be found among the complex numbers mapped out on his diagram. There was no need to extend the number system.

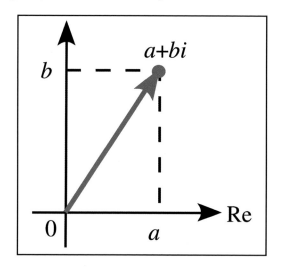

A diagram of Argand's complex number equation.

Chapter 8

PUTTING THINGS IN PERSPECTIVE

Putting things in perspective

How can we make a flat, two-dimensional image give the illusion of depth? The answer is to use linear perspective. Perspective is the art and mathematics of achieving a realistic depiction of a three-dimensional object on a two-dimensional surface. The principles of perspective drawing were first formulated in Italy in the 15th century by the Florentine architect Brunelleschi. The study of the projection of objects in a plane is called projective geometry, founded by Girard Desargues (1591–1661).

Working out perspective is the key to making two dimensions look like three.

Music timeline

5th century BC	Agatharchus of Athens invents scene painting.
13th century AD	The Italian painter Giotto creates an illusion of depth in his paintings.
*C.***1420**	Filippo Brunelleschi demonstrates his understanding of perspective.
1435	Leon Battista Alberti sets out a mathematical basis for linear perspective.
*C.***1470**	Piero della Francesca puts forward geometric theorems of perspective.
*C.***1639**	Girard Desargues founds the study of projective geometry.

THE ART OF ILLUSION

'OK, one last time. These are small... but the ones out there are far away. Small... far away...'
Father Ted, British TV comedy.

The Greeks were interested in producing realistic effects in their theatrical sets and scenery – called 'skenographia' or 'illusionism' – and techniques for achieving perspective were discussed as early as the 5th century BC by the painter Agatharchus of Athens, among others. Agatharchus even wrote a commentary on his use of convergent perspective and Greek geometers made an attempt to analyse the effect, but there is no evidence that they understood the mathematical principles behind representing perspective. A similar technique of perspective art was also employed in the mural paintings at Pompeii.

(Above) *Skenographia would almost certainly have been used in the set design employed at the Greek theatre of Epidaurus.*

But not all cultures were interested in depicting things realistically. The Egyptians, for example, disregarded perspective in their art altogether. For them, the important thing was to show figures sized according to their status in society. The Byzantine art of the 5th to 15th centuries AD similarly ignored perspective, and in Chinese painting perspective only gained importance in the 17th century.

A mosaic design from Pompei showing the use of perspective.

PERSPECTIVE AND THE RENAISSANCE

Linear perspective makes use of the fact that objects appear to get smaller the farther away they are, and that parallel lines and planes extending away from the viewer converge in the distance at a 'vanishing point'. The 13th-century painter Giotto frequently created the impression of depth by using sloping lines. Those above the eye line of the observer sloped down, while those below sloped up. Lines to the side inclined towards the centre. Geometric analysis of Giotto's work appears to show that he failed to use an accurate vanishing point and was perhaps able to achieve his effects simply because he had a good eye rather than a sound understanding of linear perspective.

An example of Giotto's attempt at perspective.

BRUNELLESCHI AND ALBERTI

The first person credited with a mathematical understanding of perspective is the Florentine architect Filippo Brunelleschi (1377–1446), best known for his work on the dome of Florence Cathedral. He understood that there should be a single vanishing point to which all parallel lines in a plane converge, although this can be – and frequently is – out of view. Also important was his understanding of scale. He correctly computed the relationship between the actual length of an object and the way its length in the picture should change, depending on its distance from the observer in the plane of the canvas. As an architect, Brunelleschi was skilled in geometry and surveying and it is reasonable to assume that he made good use of these skills in reaching his understanding of perspective.

Brunelleschi's demonstration

Around 1420, Brunelleschi made a simple but dramatic public demonstration of the power of perspective, in his native city, using the mathematical principles he had formulated. First, he painted a view of the Baptistry of San Giovanni on a small wooden panel. Then he drilled a hole, 'the size of a lentil' according to one observer, through the centre of the panel. He then invited people to stand in the piazza in front of the church and look at it through the hole in the back of the panel. Then he'd ask them to hold a mirror in front of the panel while continuing to look through the hole. Reflected back was the perfectly rendered perspective representation of the church painted on the panel, looking just like the real thing. Sadly, Brunelleschi's painting hasn't survived, but a fresco by Masaccio from this same period that uses Brunelleschi's mathematical principles still exists.

Although there is no doubt of Brunelleschi's practical mastery of perspective, he failed to set out any explanation of how his rules worked. Fellow Italian Leon Battista Alberti (1404–72) was the first to write down his ideas about linear perspective in two works aimed at different audiences. The first, entitled *De pictura*, was written in Latin

in 1435 and aimed at scholars, while the second, *Della pittura*, dedicated to Brunelleschi, was written in Italian the following year and aimed at a general audience. Alberti received his mathematical training from his businessman father and he took great pleasure in the practical application of mathematical principles. 'Nothing pleases me so much,' he wrote, 'as mathematical investigations and demonstrations, especially when I can turn them to some useful practice.'

The Baptistry of San Giovanni. Brunelleschi used a view of the baptistry to demonstrate reflective perspective to the people of Florence.

De pictura is in three parts, the first of which gives the mathematical description of perspective. Just how important Alberti reckoned a proper understanding of perspective to be can be seen in his definition of what a painting is: 'A painting,' he wrote, 'is the intersection of a visual pyramid at a given distance, with a fixed centre and a defined position of light, represented by art with lines and colours on a given surface.'

Alberti also discusses the principles of geometry and the science of optics and gives a precise

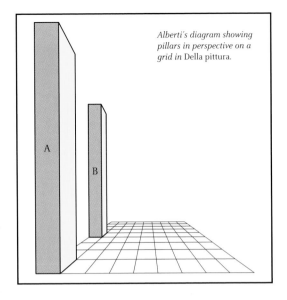

Alberti's diagram showing pillars in perspective on a grid in Della pittura.

explanation of the concept of proportionality that determines the apparent size of an object in the picture relative to its actual size and distance from the observer, which Brunelleschi knew about but failed to explain.

The idea of the visual pyramid, mentioned in the quote above, was also fully developed. The pyramid has its point of origin, or apex, in the eye of the viewer. The sides of the pyramid extend outwards from the apex following the edge of the field of vision. The painting could be imagined as a plane (a flat surface) that intersects the visual pyramid, and the apex of the pyramid as being at the ideal point for viewing the image. The vanishing point in the image, on which lines converged, was envisaged as being as far beyond the plane of the painting as the apex was in front of it. The artist was to imagine the painting as being like a window through which the observer sees the scene. A horizon runs across the canvas at eye level, and the vanishing point is located somewhere near the centre of this line. This technique is called 'one-point perspective'.

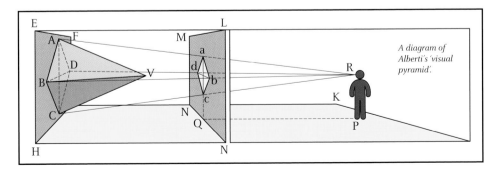

A diagram of Alberti's 'visual pyramid'.

PAVIMENTO

One of the most famous examples given by Alberti was that of a floor covered with square tiles. The vanishing point is situated in the centre of the picture and the tiles are assumed to have one edge parallel to the bottom of the picture. The sides of the tiles, which in reality are perpendicular to the bottom edge, will appear in the picture to converge on the vanishing point. The diagonals of the squares will all converge on a point positioned on a line taken through the vanishing point and running parallel to the bottom of the picture. The actual positioning of this point determines the distance the observer has to be from the picture to obtain the correct perspective effect.

Alberti didn't give mathematical proofs of his ideas, instead writing: 'We have talked as much as seems necessary about the pyramid, the triangle, the intersection. I usually explain these things to my friends with certain tedious geometrical proofs, which in this commentary it seems to me better to omit for the sake of brevity.'

Alberti's ideas were hugely influential on the development of the Renaissance style of painting. It has been said that his principles underpinned the science of perspective as much as Euclid's did in plane geometry. Pictures from this period that include a square-tiled floor are called *pavimento* (Italian for floor) pictures. There are many examples of such pictures in the years following the publication of Alberti's book.

The *pavimento* provides a type of co-ordinate system, which Alberti makes use of in showing how to use the grid to obtain the correct shape for a circle. Place a circle on a square grid and mark where the squares cut the circle. Construct the perspective view of the square grid as described above and then map on to it the points where the circle intersected the squares on the original grid. Doing so will result in the circle being projected as an ellipse.

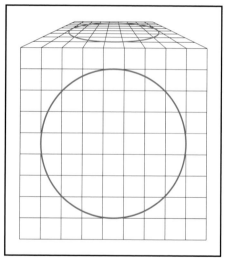

A circle projected as an ellipse.

PIERO DELLA FRANCESCA

The most mathematical account of perspective during the 15th century came from Piero della Francesca (c.1416–92), who was a leading mathematician as well as one of the greatest artists of his day. In his *On Perspective for Painting* (possibly written in the 1460s or mid-1470s), he describes the 'three principal parts' of painting – drawing, proportion and colouring. Of most interest to him was proportion, or perspective.

He begins by establishing geometric theorems in the style of Euclid, illustrating them with numerical examples. He then goes on to give theorems that relate to the perspective of plane figures and examines how to draw prisms in perspective. He developed mathematical formulae to compute the size an object should be painted on the canvas relative to its distance from the observer. He also dealt with depicting more complex objects using a method involving two rulers, one to measure width, the other height, in effect developing a co-ordinate system against which to plot the correct position of points on the object being represented and, like Alberti, anticipating Descartes by two centuries.

LEONARDO DA VINCI

Leonardo da Vinci (1452–1519) was one of the most singularly talented individuals of his, or any other, age. He developed mathematical formulae to compute the relationship between the distance from the eye to the object and its size on the canvas. He was one of the first people to study the problem of where the viewer must be situated in order to see a perspective drawing correctly. He realized that a picture painted in correct linear perspective only looked right if viewed from just the right location. This was something Brunelleschi must have been aware of when he staged his demonstration.

Leonardo recognized two different types of perspective: 'artificial perspective', which was the way the painter projects the image on to a plane, and 'natural

A sketch showing Leonardo's mastery of perspective.

perspective', which reproduces faithfully the relative size of objects depending on their distance. In natural perspective, Leonardo realized, objects will be the same size if they lie on a circle centred on the observer. He also studied compound perspective, where the natural perspective is combined with a perspective produced by viewing at an angle.

DESARGUES' THEOREM

It is a well-known law of geometry, established by Euclid, that parallel lines do not meet. However, in the world of art and perspective they *appear* to do just that at the vanishing point. French mathematician Girard Desargues (1591–1661) was one of the founders of projective geometry. His idea was to take the vanishing point and incorporate it into geometry. He called it a 'point at infinity'. Just as Euclid had set out his axioms for the study of lines on a flat plane, Desargues began a study of shapes on a plane with the addition of points at infinity. On the projective plane, there were no such things as parallel lines. In terms of projective geometry, conic sections, the circle, ellipse, parabola and hyperbola are revealed as different perspectives of the same curve. In fact, to Desargues it seemed obvious that any conic can be projected into any other conic.

Suppose you were trying to represent a flooring with triangular tiles in a painting? How could you ensure it was correctly shown in perspective? Desargues came up with a solution to that problem. In what became known as Desargues' theorem, he stated that if two triangles are in perspective from a point, then they're also in perspective from a line. This means that if you extend lines joining the corresponding vertices of each triangle (the actual triangle and the perspective rendition of it) they should all meet at a single point. Also, if you extend the edge of one triangle and the corresponding edge of the other they should meet at a single point. Doing this with all three edges produces three points that should lie along a straight line. If these requirements are met, the triangles are in perspective. Consideration of Desargues' theorem has been of huge practical benefit to artists ever since.

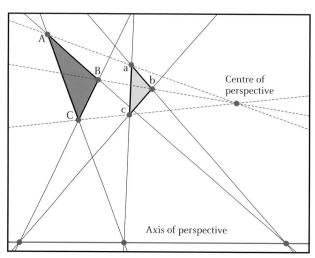

Desargues' theorem.

Chapter 9

PROBABILITY – WHAT ARE THE CHANCES?

Probability – what are the chances?

The mathematics of probability are concerned with solving the problem of predicting seemingly random events. The people of ancient Greece, Rome and India loved games of chance, but no one that we know of made any attempt to understand or to discover the mathematical laws that lay behind chance events. Luck was simply something that happened at the whim of the gods.

It wasn't until the 16th century that the first attempts were made at predicting outcomes. Gerolamo Cardano (see also page 64) was a compulsive gambler as well as a mathematician, and, like any gambler, he was interested in finding a way to increase his likelihood of winning. Solving this problem led him to the first scientific analysis of probability as he sought a way of attributing numerical values to chance events. This revolutionary idea resulted in probability theory, which in turn led to statistics, without which we wouldn't have such things as the insurance industry and the weather forecast.

Fresco of dice players in Pompei.

Probability timeline

c. 1564 — Gerolamo Cardano writes *Liber de ludo aleae* (Book on Games of Chance), the first systematic approach to probability.

c. 1650 — Blaise Pascal and Pierre de Fermat lay the foundations of modern probability theory.

1657 — Christian Huygens publishes the first book on probability, *De ratiociniis in ludo aleae* (On Reasoning in Games of Chance).

1662 — John Graunt uses statistics to predict life expectancy, the beginning of the science used to set insurance premiums.

1763 — Thomas Bayes' work, 'An Essay Towards Solving a Problem in the Doctrine of Chances', introduces conditional probability.

1812 — Pierre de Laplace, in his book *Théorie analytique des probabilités*, extends probability theory beyond games of chance and applies it to scientific and other practical problems.

FIRST STEPS

Gerolamo Cardano was born in Pavia, in present-day Italy, in 1501. Educated at the universities of Pavia and Padua, he practised as a medical doctor from 1526 to 1553, during which time he also studied mathematics and other sciences. He published several works on medicine and in 1545 published a text on algebra, the *Ars magna*, which was hugely influential (see page 65).

A chess player as well as an accomplished gambler, among Cardano's other works is the *Liber de ludo aleae* (Book on Games of Chance) started when he was just 25 years old,

but unpublished until 1663, long after his death. It contains perhaps the first systematic treatment of probability (as well as a section on effective cheating methods).

Some of its ideas may seem obvious to us now, but Cardano's book was ahead of its time. The notion that where there are several equally likely outcomes the chance of any one event occurring is the same as all the other possible outcomes hadn't been expressed before. For example, when you throw a die each of the six sides has the same chance of coming up, so the chance of any one side coming up – '5', for example – is one in six. And that is true no matter how many times '5' has already appeared.

Portrait of Girolamo Cardano.

THE UNFINISHED GAME

The fundamental principles of probability theory were formulated through the exchange of a series of letters between Blaise Pascal and Pierre de Fermat. The correspondence came about as a result of a gambling dispute.

Imagine two people are playing a game of chance with money at stake. A point is awarded to one of the players, based on how a coin lands, one winning on heads and the other on tails. The first player to make 10 points wins the money. For some reason, the game is interrupted and can't be continued. At this stage one player has 8 points and the other has 7. How should they divide the pot? Player One may be ahead on points, but Player Two still has a good chance of winning. This problem is often referred to as the problem of points.

The answer Pascal and Fermat came up with involved using a method of computing probabilities known as 'enumeration'. This requires enumerating (or listing) all possible outcomes of an action, like a series of coin tosses. If we decide which of the outcomes denotes success, then by adding together all the outcomes that we have decided are successful we arrive at the probability of success.

In the interrupted game, how many more tosses would be required to decide a winner? It could be as few as two for Player One to win or a maximum of four if both players make it to 9 points. So, we draw up a table, Fermat called it a table of 'future possibilities', setting out all the possible results from four tosses of the coin. There are 16 of them.

H H H H	H T H H	T H H H	T T H H
H H H T	H T H T	T H H T	T T H T
H H T H	H T T H	T H T H	T T T H
H H T T	H T T T	T H T T	T T T T

All of the possible outcomes in light grey are those in which Player One wins – 11 out of the 16. The probability of Player One winning the game is $^{11}/_{16}$ and therefore, concluded Fermat, Player One should be awarded $^{11}/_{16}$ths of the pot and Player Two the remaining $^{5}/_{16}$ths, fairly reflecting his chances of winning. You'll doubtless have noticed that in a few instances some of the tosses are redundant, the game having been won in the first two tosses, but all of the sequences have to be completed to make each one equally probable and thus comparable with the others.

The coin toss illustrates the theory of probability.

PROBABLE OUTCOMES

We can use this approach to work out the probability (P) of any event occurring. P = the number of ways an event can occur divided by the number of possible events. Because the number of occurrences can never exceed the number of possibilities, P is always somewhere on a scale of 1 to zero. If an event is impossible – rolling a '7' with one die, say – the probability is zero. If it is certain to happen – rolling any number between '1' and '6' for instance – the probability is 1. All other possible events are somewhere in between.

If we want to find the probability of two mutually exclusive events occurring – for example, the probability of throwing either a '4' or a '5' with one die – we simply add up the probability of each event, which in this case is $^{1}/_{6} + ^{1}/_{6} = ^{1}/_{3}$.

To work out the probability of multiple independent events occurring, we have to multiply the probabilities together. For example, the chances of throwing two sixes in a row are $^{1}/_{6} \times ^{1}/_{6} = ^{1}/_{36}$.

PASCAL'S TRIANGLE

Pascal's approach was slightly different but it came to the same conclusion. He was looking for a way of solving the problem that avoided having to list all the possible outcomes. He realized that he could generate the numbers using a triangle of numbers he had devised.

Pascal's triangle is a table of binomial coefficients. A binomial is an expression with two terms that are operated on by a straightforward arithmetic operation, such as multiplying, dividing, adding or subtracting. A coefficient is a number that multiplies a variable – for example, in the expression $3x + 2 = y$, 3 is the coefficient of the variable x.

Pascal was not the first to devise the triangle that bears his name. The first recorded occurrence is in the work of Pingala, an Indian writer of around 200BC, and later, in 1303, by the Chinese mathematician Szu Yuen Yu Chien. Putting the triangle together is simple. Begin with a single 1 at the apex. Below it, write two more 1s. Each subsequent row begins and ends with 1 and each number in between is calculated by adding together the values of the numbers above. As the triangle is extended, patterns begin to appear in the numbers.

To see how Pascal's triangle can be used to solve the problem of points, recall that the maximum number of coin tosses required to finish the interrupted game is four. The top row, the '1' at the apex, is known as row 'zero', not row 'one'. If we construct the triangle down as far as row 'four' this is what we get:

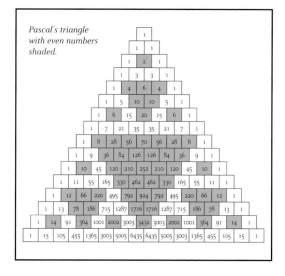

Pascal's triangle with even numbers shaded.

Row zero	1	
Row one	1 1	
Row two	1 2 1	
Row three	1 3 3 1	
Row four	1 4 6 4 1	

Row 'four' corresponds to the number of possible combinations from four coin tosses:
- 1 way of getting four heads.
- 4 ways of getting three heads and a tail.
- 6 ways of getting two heads and two tails.
- 4 ways of getting three tails and a head.
- 1 way of getting four tails.

As we know Player One only needs two heads to win the game, all we have to do is add up all the combinations where that happens – $1 + 4 + 6 = 11$, or an $^{11}/_{16}$ chance of winning, which is, of course, the same answer as Fermat reached.

Other mathematicians were soon following up and expanding on Fermat and Pascal's insights. In 1657, Dutch scientist and mathematician Christiaan Huygens became the first to publish a text on probability theory entitled *De ratiociniis in ludo aleae* (On Reasoning in Games of Chance) in which he introduced the idea of 'mathematical expectation', or the expected value, a means of using probability to determine what outcomes we can expect over the long term. He acknowledged the Pascal–Fermat letters as his inspiration.

Then came Jacob Bernoulli (1654–1705), eldest of a family of renowned Swiss mathematicians, whose *Ars conjectandi* (The Art of Conjecture, published posthumously in 1713) was the first to use the word 'probability' in its modern sense. Bernoulli provided detailed explanations of Huygens' proofs and offered alternatives of his own, including tackling a series of problems on games of chance. Perhaps most importantly, he unveiled what became known as the 'Bernoulli theorem', and later called the 'law of large numbers' (see box, page 85).

Frenchman Abraham de Moivre continued the progress of probability. He was one of the first to try to come to terms with the deep mathematics underpinning the theory. His most important contribution was his demonstration that the distribution of natural phenomena frequently averaged out into a bell-shaped curve, or 'bell curve', when shown on a graph, although he failed to recognize its importance at the time. Later it would be named the 'normal distribution' by the great mathematician Carl Gauss.

Pierre de Fermat.

THE CENTRAL LIMIT THEOREM

The normal distribution comes up time and time again, even in places where you might not expect to find it. For example, tossing coins is an activity with only two possible outcomes – heads or tails. How can this form a bell curve? De Moivre found the first hints of what would later come to be known as the 'central limit theorem'. Flip a coin a hundred times and record the number of heads. Now flip another 99 series of a hundred tosses and record the results for each of those. (I'm sure you're glad you don't actually have to do this.) Plot the results on a graph and we find that results are distributed around the average of 50 that we would have expected. The more times the sequence is carried out the closer the results come to following the 'normal distribution' curve.

The central limit theorem states that, given a sufficiently large sample size from a population with a finite level of variance (heads or tails; red, white or blue), the mean of a sample of data will be closer to the mean of the overall population as the sample size increases and that all the samples will approximate a 'normal distribution' pattern. As a general rule, sample sizes of 30 or more are considered sufficient for the central limit theorem to hold.

The theorem is used in fields such as finance, where investors use it to analyse stock returns and construct their portfolios. For instance, an investor looking at the performance of a stock index of 1,000 stocks could sample 30 of them at random and construct a distribution pattern that would reflect the performance of the index as a whole.

The normal distribution

The 'normal distribution' has gone on to take its place across the whole of science. It is determined by two numbers: the expected, or 'mean', value and the 'standard deviation', which is a measure of how spread out the numbers are around the mean. For example, were you to measure the heights of 100 randomly chosen people and plot the results on a graph, you will produce a bell-shaped standard distribution. You would find the majority of the sample clustered around the same height: this is the mean height of the group. The further away from the mean in either direction, the fewer examples there will be. The greater the variation from the mean, the larger the standard deviation; the larger the standard deviation, statistically speaking, the less likely it is there will be an example that fits that point on the graph.

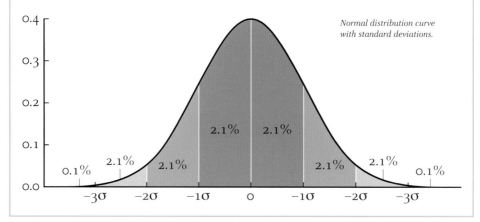

Normal distribution curve with standard deviations.

The law of large numbers

The law of large numbers was first proved by the Swiss mathematician Jacob Bernoulli in 1713. It is a probability theorem that states that as the number of identically distributed, randomly generated variables increases, their observed average approaches their theoretical average. In other words, if you toss a coin a sufficiently large number of times you should tend to get an even distribution of heads and tails. It took Bernoulli twenty years to develop his mathematical proof that this was so. More than a century later, the Russian mathematician Pafnuty Chebyshev (1821–94) demonstrated that the law of large numbers is closely related to what is commonly known as the law of averages. When tossing a coin, the law of large numbers indicates that the number of heads tossed will come closer and closer to half the total – a probability of 0.5 – though it can take many tosses for this to happen. To obtain a 95 per cent probability of the fraction of heads falling between 0.47 and 0.53 requires in excess of 1,000 tosses.

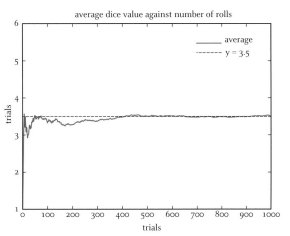

The more trials that are made, the closer the actual result gets to the theoretical result.

THE BIRTH OF STATISTICS

In 1662, businessman John Graunt published a book called *Natural and Political Observations Made Upon the Bills of Mortality*. This is considered the founding text of statistics and also of demography – the use of statistics to analyse human populations, including births, deaths and the incidence of disease. The book was an analysis of fifty years of data extracted from the Bills of Mortality, the weekly mortality listings for London, that set out how many people had died in the previous week, how they had died, and who they were. In his efforts to make sense of the bills, Graunt produced the first known tables of health data. Based on this information, Graunt made the first realistic estimate of London's population, so the book may also be said to mark the beginnings of population statistics.

Death comes to individuals in unpredictable ways, but Graunt discovered that where large populations were involved the number of people dying of particular causes became much more regular and predictable. Graunt's analyses gave him insights into possible

The Difeafes, and Cafualties this year being 1632.

Abortive, and Stilborn	445	Jaundies	43
Affrighted	1	Jawfaln	8
Aged	628	Impoftume	74
Ague	43	Kil'd by feveral accidents	46
Apoplex, and Meagrom	17	King's Evil	38
Bit with a mad dog	1	Lethargie	2
Bleeding	3	Livergrown	87
Bloody flux, fcowring, and flux	348	Lunatique	5
Brufed, Iffues, fores, and ulcers,	28	Made away themfelves	15
Burnt, and Scalded	5	Meafles	80
Burft, and Rupture	9	Murthered	7
Cancer, and Wolf	10	Over-laid, and ftarved at nurfe	7
Canker	1	Palfie	25
Childbed	171	Piles	1
Chrifomes, and Infants	2268	Plague	8
Cold, and Cough	55	Planet	13
Colick, Stone, and Strangury	56	Pleurifie, and Spleen	36
Confumption	1797	Purples, and fpotted Feaver	38
Convulfion	241	Quinfie	7
Cut of the Stone	5	Rifing of the Lights	98
Dead in the ftreet, and ftarved	6	Sciatica	1
Dropfie, and Swelling	267	Scurvey, and Itch	9
Drowned	34	Suddenly	62
Executed, and preft to death	18	Surfet	86
Falling Sicknefs	7	Swine Pox	6
Fever	1108	Teeth	470
Fiftula	13	Thrufh, and Sore mouth	40
Flocks, and fmall Pox	531	Tympany	13
French Pox	12	Tiffick	34
Gangrene	5	Vomiting	1
Gout	4	Worms	27
Grief	11		

Chriftened { Males—4994 Females—4590 In all —9584 } Buried { Males —4932 Females —4603 In all —9535 } Whereof, of the Plague;8

Increafed in the Burials in the 122 Parifhes, and at the Pefthoufe this year 293
Decreafed of the Plague in the 122 Parifhes, and at the Pefthoufe this year, 266

C   7 In

A page from John Graunt's Natural and Political Observations.

origins of the plague that had been devastating the population. He observed that other causes of death, such as tuberculosis, which killed around 2,000 people a year at the time, followed a regular pattern. But plague was different. It claimed 46,000 lives in 1625 and none at all four years later. The contrast between the irregularity of plague deaths and the regularity of deaths caused by chronic diseases suggested to Graunt that plague had an environmental origin.

Graunt's work was another source of inspiration for Huygens, who was sent a copy of Graunt's book. Using Graunt's life tables, Huygens and his brother, Lodewijk, worked together on ideas for predicting life expectancy. In 1693, Sir Edmund Halley (perhaps better known for the comet named after him) used Graunt's ideas to create the first actuarial tables for the nascent life insurance industry. He was asked by the Royal Society to examine records of births and deaths from the German city of Breslau. From these Halley produced his own life tables, and from them he derived a formula that showed how to use the chances of death at any given age to calculate the cost of an annuity.

THE AVERAGE PERSON

In 1835, Adolphe Quetelet (1796–1874), a Belgian mathematician, astronomer and statistician, combined statistics and probability to reveal his idea of *l'homme moyen* (the average man). The 'average man' represented the central value around which human characteristics are grouped according to the normal distribution. His pioneering studies of human growth led him to conclude that, spurts of growth following birth and during puberty aside, 'weight increases as the square of the height'. This was known as the Quetelet Index until 1972 when it was renamed the Body Mass Index (BMI).

Quetelet's concept is applied today to plan for public health concerns. In his original study, he measured the heights of 100,000 army conscripts and compared the actual data with the expected value. He was puzzled to find that the numbers of people just below and above the height limits for military service were greater than expected. As Quetelet was able to rule out measurement errors, he concluded that the conscripts had been lying to avoid conscription.

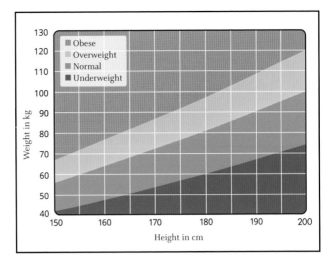

Body mass index.

REAL LIFE PROBLEMS AND PROBABILITIES

It's all very well to consider the probabilities involved with tossing coins, throwing dice, playing cards and whether or not to buy a lottery ticket. But real life is much messier than any of those. How can probability help with the problem of predicting what might happen there?

In order to make probability a useful tool for making real world predictions, a way had to be found to take in events in which the outcome was not fixed – not simply heads or tails, whatever else happened – but was tied up with the outcomes of other events. The answer came with the development of conditional probability.

CONDITIONAL PROBABILITY

Throw a die and you know you're going to get one of six numbers. It's only while the die is rolling that there's any uncertainty about which number you will get – once it stops you're 100 per cent certain that you've got a '3', or whatever. But why did you? Did something intervene in some way that altered the expected one-in-six chance and make it more likely that you'd get a '3' rather than the '6' you were hoping for?

In the real world, we're used to events affecting other events in seemingly unpredictable ways. We might call it luck, good or bad, fate, the answer to a prayer, or any number of other terms. Or maybe things just happen, as Harold Macmillan reputedly put it – 'Events, dear boy, events'. The breakthrough in mathematically modelling events affecting each other came in the early 18th century.

In 1763, a paper was published entitled 'Essay towards solving a Problem in the Doctrine of Chances'. It was a landmark in mathematical problem solving. The author was the Reverend Thomas Bayes, who had died two years earlier.

Bayes considered two events, call them 'Event A' and 'Event B'. Each has a probability of occurring, P(A) and P(B), with P for each being, as we have seen, a number between 0 and 1. What Bayes did was to make the events interdependent. If Event A occurred it would alter the probability of Event B happening, and vice versa. A might guarantee that B will take place or it might prevent it from happening. To denote this, Bayes introduced two new quantities, now referred to as conditional probabilities. These are given as P(A|B), the probability of A given B, and P(B|A), the probability of B given A. What Bayes did was to solve the problem of how all four probabilities related to one another. In the theorem that bears his name he gave the answer:

$P(A|B) = P(A) \times P(B|A)/ P(B)$.

So what does it mean? Interpreting the results of medical tests offers a good example of how Bayes' theorem works in the real world. Suppose a patient shows symptoms that could indicate a serious but rare illness (I) that affects 1 per cent of the population. There is a very good test (T) for the illness that has a 95 per cent reliability. This means that 95 out of 100 people who have the illness will test positive, but it also means that 5 out of 100 who are actually ill will test negative (a false negative) and 5 out of 100 who are in fact healthy will test positive (a false positive). What does a positive test mean in this case? If you assume it means the patient has a 95 per cent chance of having the illness, you'd be wrong. Let's plug the figures into Bayes' equation to find out why.

P(I) the probability of the patient having the illness is 0.01

P(I|T) the probability that the patient has the illness and tests positive is 0.95

P(T) the probability that the test says the illness is present, whether or not it actually is, is found by multiplying the rate of false positives (0.05) by the percentage of the population who don't get the illness (0.99), which is 0.0495. To this we add the false negatives, calculated in the same way, to arrive at a total of 0.099

P(T|I) the probability that the patient has the illness and tests positive is what we have to determine. Putting it all together we have:

$P(T|I) = 0.01 \times 0.95/0.099$

$= 0.0959$

In other words, this is just under a 10 per cent chance of having the illness, even if the test is positive. Perhaps somewhat lower than you may have thought!

Bayes theorem is a powerful tool with many uses, but it must be used wisely. Like anything else, it is only as good as the evidence used to supply the figures. Sound figures will produce solid results. Poor figures are useless.

Chapter 10

LOGARITHMS – CALCULATING MADE EASY

Logarithms – calculating made easy

When John Napier published his tables of logarithms in 1614 it had an enormous impact. It revolutionized the way people carried out calculations, including mariners for navigation, land and military surveyors for their plans and particularly astronomers, who often had to calculate using very large numbers. The French mathematician Pierre-Simon Laplace (1749–1827) remarked gratefully that Napier's new tool 'doubled the life of the astronomer'.

Napier also worked on other calculating methods, finding ways to speed up the work of calculating his logarithms. He was the first to introduce the use of the decimal point and to propose using binary numbers in calculations. He also invented Napier's bones – a series of rods that could be assembled in different ways to multiply or divide large numbers by reading across the columns of figures. Isaac Newton used them to aid his calculations. Logarithms remained essential tools for scientists and engineers until the advent of the electronic calculator in the 1970s.

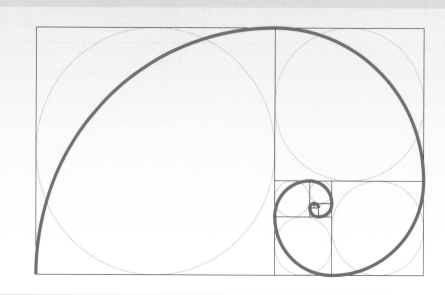

An example of a logarithmic spiral.

Logarithmic timeline

1484   Nicolas Chuquet writes about geometric series in *Triparty*.

1614   John Napier publishes his work on logarithms in *Mirifici logarithmorum canonis descriptio* (A Description of the Marvellous Rule of Logarithms).

1617   Napier produces his calculating device – Napier's bones.

1620   Edmund Gunter makes a straight logarithmic scale and performs multiplication and division on it using a set of dividers.

1622   William Oughtred invents the slide rule using logarithmic scales.

1624   Henry Briggs publishes his improved recalculation of Napier's logarithms.

1792   Gaspard de Prony begins the task of producing the *Cadastre*, a set of logarithmic and trigonometric tables given to between 14 and 29 decimal places.

SIMPLIFYING CALCULATIONS

Science was flourishing in the late 16th century with new developments in many fields such as navigation and astronomy. For many scientists, there were long and laborious calculations to be performed, all by hand. Reducing this arduous burden, and the errors that inevitably crept in to long calculations, was a problem that had to be solved. Great effort was put into finding new computational techniques.

One avenue of interest was in exploring techniques that bypassed the need for multiplication and division by replacing these processes with the much more straightforward addition and subtraction. One method originating in the late 16th century transformed long multiplication and division into addition and subtraction by using trigonometry tables

In the 15th and 16th centuries, mathematicians such as Nicolas Chuquet (c.1445–88) and Michael Stifel (c.1487–1567) turned their attention to the relationship between arithmetic and geometric series as a path towards uncovering more powerful methods of computation. An arithmetic series is a sequence of numbers in which each number differs from the one preceding it by a constant quantity, such as 1, 2, 3, 4, 5, 6... and so on, or 2, 4, 6, 8...

A geometric series is one in which each number after the first term is found by multiplying the previous number by a fixed amount, called the 'common ratio'. For example, the sequence 16, 8, 4, 2, 1 has a common ratio of $\frac{1}{2}$. Here's one way that combining an arithmetic and a geometric series can simplify calculation:

0	1	2	3	4	5	6	7	8	9	10
1	2	4	8	16	32	64	128	256	512	1024

The bottom row of the table, which is a geometric series, shows powers of two; the top row, which is an arithmetic series, shows the exponents by which two must be raised to arrive at the powers. In other words, the exponent indicates how many times the number must be multiplied by itself to give the figure shown. For example, 2^4 (that is, 2 × 2 × 2 × 2) = 16. You might be able to multiply 8 by 128 in your head, but an easier way to do it is to look at the table. When multiplying exponents of the same base we simply add the exponents together: 8 multiplied by 128 is the same as 2^3 multiplied by 2^7, which is 2^{3+7} or 2^{10}, which is, from the table, 1024. I'm sure you'll agree that adding 3+7 is a lot easier than multiplying 8 by 128. It was a much more sophisticated version of this very technique that was at the heart of Napier's logarithms.

John Napier

John Napier (1550–1617), known as the 8th Laird of Merchiston, was a Scottish aristocrat who lived in what is now part of Napier University in Edinburgh. He was educated at St Andrew's University, entering the university in 1563 at the age of 13, but left before completing his degree. It is likely that he went on to study in Europe, but no records exist of where he might have done so.

As well as his interest in mathematics, Napier was also a keen student of theology and of armaments. There is a document bearing his signature in which a variety of inventions are listed, including a metal chariot with small holes through which weapons could be discharged – a sort of very early tank.

He cultivated an eccentric manner, seldom being seen outdoors without his trademark black cloak and black rooster. He had a reputation for being a wizard. According to one story, he exposed a thief among his staff by asking them one by one to stroke his rooster (which he had surreptitiously blackened with soot). The magic chicken, he said, would mark the guilty party's hand. The innocent members of staff, with nothing to hide, were happy to stroke the bird, while the culprit did not, and was revealed by his clean hands.

John Napier.

The basic idea behind logarithms was simple enough: it was to replace the often long and wearisome task of multiplying two numbers by the simpler task of adding together two other numbers. Each number was to have its equivalent 'artificial number' (Napier later settled on the name 'logarithm', which he coined by combining two Greek terms, *logos*, meaning proportion, and *arithmos*, meaning number) such that the product

II. Tabula Logarithmorum

5 GRAD.

M.|S.	Log. Sinus	Diff.1"	Log. Cofin.	D.1"	Log. Tang.	C.D.1"	Log. Cot.	"|'
24 0	8.973 6280	222.7	9.998 0683	2.0	8.975 5597	224.7	11.024 4403	0 |56
10	8.973 8507	222.5	9.998 0663	2.0	8.975 7844	224.5	11.024 2156	50
20	8.974 0732	222.5	9.998 0643	2.0	8.976 0089	224.5	11.023 9911	40
30	8.974 2957	222.3	9.998 0623	2.0	8.976 2334	224.3	11.023 7666	30
40	8.974 5180	222.3	9.998 0603	2.0	8.976 4577	224.3	11.023 5423	20
50	8.974 7403	222.1	9.998 0583	2.0	8.976 6819	224.1	11.023 3181	10
25 0	8.974 9624	222.0	9.998 0563	2.0	8.976 9060	224.0	11.023 0940	0 |35
10	8.975 1844	221.8	9.998 0543	2.0	8.977 1300	223.9	11.022 8700	50
20	8.975 4062	221.8	9.998 0523	2.0	8.977 3539	223.8	11.022 6461	40
30	8.975 6280	221.7	9.998 0503	2.0	8.977 5777	223.6	11.022 4223	30
40	8.975 8497	221.5	9.998 0483	2.0	8.977 8013	223.5	11.022 1987	20
50	8.976 0711	221.4	9.998 0463	2.0	8.978 0248	223.5	11.021 9752	10
26 0	8.976 2926	221.3	9.998 0443	2.0	8.978 2483	223.3	11.021 7517	0 |34
10	8.976 5139	221.2	9.998 0423	2.0	8.978 4716	223.2	11.021 5284	50
20	8.976 7351	221.1	9.998 0403	2.0	8.978 6948	223.1	11.021 3052	40
30	8.976 9562	221.0	9.998 0383	2.0	8.978 9179	222.9	11.021 0821	30
40	8.977 1772	220.8	9.998 0363	2.0	8.979 1408	222.9	11.020 8592	20
50	8.977 3980	220.8	9.998 0343	2.0	8.979 3637	222.8	11.020 6363	10
27 0	8.977 6188	220.6	9.998 0323	2.0	8.979 5865	222.6	11.020 4135	0 |33
10	8.977 8394	220.5	9.998 0303	2.0	8.979 8091	222.5	11.020 1909	50
20	8.978 0599	220.4	9.998 0283	2.0	8.980 0316	222.4	11.019 9684	40
30	8.978 2803	220.2	9.998 0263	2.0	8.980 2540	222.3	11.019 7460	30
40	8.978 5006	220.2	9.998 0245	2.1	8.980 4763	222.2	11.019 5237	20
50	8.978 7208	220.0	9.998 0222	2.0	8.980 6985	222.1	11.019 3015	10
28 0	8.978 9408	220.0	9.998 0202	2.0	8.980 9206	222.0	11.019 0794	0 |32
10	8.979 1608	219.8	9.998 0182	2.0	8.981 1426	221.8	11.018 8574	50
20	8.979 3806	219.8	9.998 0162	2.0	8.981 3644	221.8	11.018 6356	40
30	8.979 6004	219.6	9.998 0142	2.0	8.981 5862	221.6	11.018 4138	30
40	8.979 8200	219.5	9.998 0122	2.1	8.981 8078	221.5	11.018 1922	20
50	8.980 0395	219.4	9.998 0101	2.0	8.982 0293	221.4	11.017 9707	10
29 0	8.980 2589	219.2	9.998 0081	2.0	8.982 2507	221.3	11.017 7493	0 |31
10	8.980 4781	219.2	9.998 0061	2.0	8.982 4720	221.2	11.017 5280	50
20	8.980 6973	219.1	9.998 0041	2.0	8.982 6932	221.1	11.017 3068	40
30	8.980 9164	218.9	9.998 0021	2.1	8.982 9143	221.0	11.017 0857	30
40	8.981 1353	218.8	9.998 0000	2.0	8.983 1353	220.8	11.016 8647	20
50	8.981 3541	218.8	9.997 9980	2.0	8.983 3561	220.8	11.016 6439	10
30 0	8.981 5729	218.6	9.997 9960	2.1	8.983 5769	220.6	11.016 4231	0 |30
10	8.981 7915	218.5	9.997 9939	2.0	8.983 7975	220.6	11.016 2025	50
20	8.982 0100	218.4	9.997 9919	2.0	8.984 0181	220.5	11.015 9819	40
30	8.982 2284	218.2	9.997 9899	2.0	8.984 2385	220.3	11.015 7615	30
40	8.982 4466	218.2	9.997 9879	2.1	8.984 4588	220.2	11.015 5412	20
50	8.982 6648	218.1	9.997 9858	2.0	8.984 6790	220.1	11.015 3210	10
31 0	8.982 8829	217.9	9.997 9838	2.0	8.984 8991	220.0	11.015 1009	0 |29
10	8.983 1008	217.9	9.997 9818	2.1	8.985 1191	219.8	11.014 8809	50
20	8.983 3187	217.7	9.997 9797	2.0	8.985 3389	219.8	11.014 6611	40
30	8.983 5364	217.6	9.997 9777	2.0	8.985 5587	219.6	11.014 4413	30
40	8.983 7540	217.5	9.997 9757	2.1	8.985 7783	219.6	11.014 2217	20
50	8.983 9715	217.4	9.997 9736	2.0	8.985 9979	219.4	11.014 0021	10
32 0	8.984 1889	217.4	9.997 9716	2.0	8.986 2173	219.4	11.013 7827	0 |28
'|"	Log. Cofin.	Diff.1"	Log. Sinus	D.1"	Log. Cot.	C.D.1"	Log. Taug.	S.|M.

84 GRAD.

An extract from a book of log tables.

of adding two artificial numbers, when converted back to an ordinary number, would yield the result of multiplying the original numbers. For division, one logarithm is subtracted from another and the result converted back.

Napier first published his work on logarithms in 1614 under the title *Mirifici logarithmorum canonis descriptio* (A Description of the Marvellous Rule of Logarithms). He was aware that most people who had lengthy calculations to deal with generally did them in the context of trigonometry and he therefore set his logarithms in a trigonometric context to make them more relevant and useful.

Napier's approach to generating his logarithms was an interesting one, and mathematicians are still not quite certain why he did it in this way. He imagined two particles travelling along two parallel lines. The first line was of infinite length while the second was of fixed length. Napier imagined the two particles leaving the same starting position at the same time with the same velocity. The first particle, on the infinite line, he set in uniform motion so that it covered equal distances in equal times. The second particle, on the fixed length, he set in motion so that its velocity was proportional to the distance remaining from the particle to the end of the line, which meant that it was slowing down. When it reaches the half-way point between the starting point and the end of the line the second particle is travelling at half the velocity it started with; at the three-quarter point, it is travelling with a quarter of the velocity; and so on. This means that the second particle is never actually going to reach the end of the line, any more than the first particle, on its infinite line, will ever arrive at the end of its journey.

At any instant on their journeys, there is a unique correspondence between the positions of the two particles. The distance of the particle on the infinite line from its starting point is the logarithm of the distance the second particle has to go to reach the end of its line. In other words, the distance the first particle has travelled at any instant is the logarithm of the distance the second particle has yet to go. Another way of looking at it, linking back to an idea we saw earlier, is that the first particle's progress is an arithmetic progression, while the second particle's is geometric.

Napier's bones

In addition to logarithms, John Napier also invented an ingenious calculating device. It consisted of a set of ten rods that could be used to carry out a variety of operations, such as multiplication, division and finding square and cube roots. The faces of the rods had a number from 0 to 9 inscribed at the top, with increasing multiples of that number along the rod from top to bottom. When the rods were placed next to each other, the products of multiplication sums could be read off, although the user still had to add up the digits for the partial products. For example, to multiply 826 by 742 the rods for 8, 2, and 6 were aligned and the user looked horizontally across the times 7, times 4 and times 2 rows to find the partial products, which were then added together.

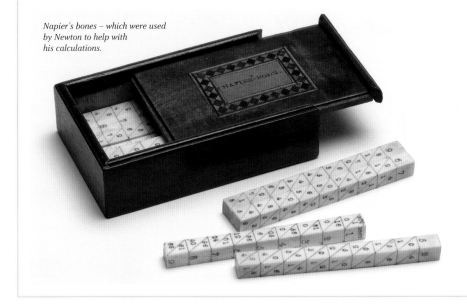

Napier's bones – which were used by Newton to help with his calculations.

Napier's original logarithms differed substantially from the form that was later adopted. Because he had intended them to be used for trigonometry, it was the logarithms of sines and tangents that he had calculated and not the logarithms of numbers in general. Napier worked with geometry professor Henry Briggs (1561–1630) on simplifying his

The Eiffel Tower in Paris – Gaspard de Prony is one of 72 names inscribed on the tower.

logarithms. It was as a result of their discussions that it was decided to redefine the logarithm of 1 as 0 and the logarithm of 10 as 1 – Napier had previously used a much more cumbersome method.

The approach agreed with Briggs made the use of logarithms much easier. Briggs also helped with the calculation of logarithms of ordinary numbers based on log 10 = 1, log 100 = 2, log 1,000 = 3, and so on and spent several years recalculating the tables on this basis. Briggs' recalculated tables were published in 1624 with the logarithms calculated to 14 decimal places. These base 10 logarithms calculated by Briggs are known as $\log_{10}$ or common logarithms. The example above of the powers of two can be thought of as a simple base 2, or $\log_2$ table.

Mathematically speaking, logarithms are simply the reverse of exponentials (powers of a number) and can be to any base.

It wasn't long before knowledge of logarithms began to spread. A translation into English from Napier's original Latin text was made by navigator Edward Wright. Briggs gave lectures on them at Gresham College in London where he was professor, and so too did Edmund Gunter, who was Professor of Astronomy at the college. Within a few years logarithm tables had been published in France, Germany and the Netherlands. Towards the end of the 18th century, Gaspard de Prony (1755–1839) oversaw the compilation of the *Tables de Cadastre*. This amazing undertaking ran to 17 volumes and included logarithms for numbers up to 200,000, all to at least 19 decimal places.

The slide rule

Before pocket calculators became ubiquitous in the 1980s (and of course now everyone has a calculator app on their smart phone) every scientist, engineer, architect and high school student would know how to use a slide rule, a calculating tool that dates back nearly 400 years.

It had its beginnings in 1620 when Edmund Gunter (1581–1626) realized that by engraving a scale of logarithms on a piece of wood and then using a pair of compasses to add together two values, he

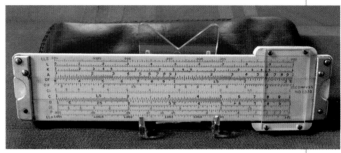

An example of a slide rule.

could eliminate the time-consuming process of looking up the logarithms in a table. William Oughtred (1574–1660), one of the leading mathematicians of his day, quickly came up with the idea that if he had two such scales marked along the edges of the pieces of wood, he could slide them relative to each other and thus do away with the need for a pair of dividers. The slide rule was born. Later inventors, including Sir Isaac Newton and the engineer and steam pioneer James Watt, came up with further improvements as the slide rule was refined and developed over the next few hundred years.

GULIELMUS OUGHTRED *Anglus* ex

William Oughtred.

THE LOGARITHMIC SCALE

A logarithmic scale is a scale of measurement that uses the logarithm of a value instead of the actual value of whatever it is that's being measured. On a logarithmic scale, each step on the scale is a multiple of the preceding step. So $\log_{10} 1 = 0$, $\log_{10} 10 = 1$, $\log_{10} 100 = 2$, $\log_{10} 1{,}000 = 3$, $\log_{10} 10{,}000 = 4$, and so on. While the numbers on a regular scale are evenly spaced, on a logarithmic scale they get closer together the further up the scale you go. Every unit up the scale represents a tenfold increase in the value of what's being measured.

For example, the decibel scale used for measuring sound levels is logarithmic rather than linear. The softest sound detectable by a normal human ear has a pressure variation of 20 micro-Pascals, which is 20×10^5 Pa. This is called the Threshold of Hearing. On the other hand, the sound pressure close to an extremely noisy event, such as a rocket leaving the launch pad, can produce a large pressure variation of approximately 2,000 Pa or 2×10^9 μPa. Expressing the magnitude of sound across a scale as wide as this, ranging from 20 to 2,000,000,000 is unwieldy and inconvenient. To avoid this problem, the decibel or dB scale is used. The decibel scale takes the hearing threshold of 20 μPa, which is defined as 0 dB as its reference level. A sound 10 times louder is assigned a decibel value of 10; a sound 100 times louder has a decibel value of 20; a sound 1,000 times louder a value of 30, and so on.

Another reason for using the logarithmic decibel scale is that, roughly speaking, a sound has to be ten times the intensity to sound twice as loud to the human ear, so it fits in well with the way we actually hear things.

Other examples of logarithmic scales used in science include the Richter Scale used to measure earth-quake intensity – an earthquake registering 7 on the scale would be ten times more powerful than one registering 6; the pH scale used to measure levels of acidity and alkalinity is logarithmic; Google's PageRank system is also logarithmic – a site with a PageRank of '5' is a hundred times more popular than one ranked '3'.

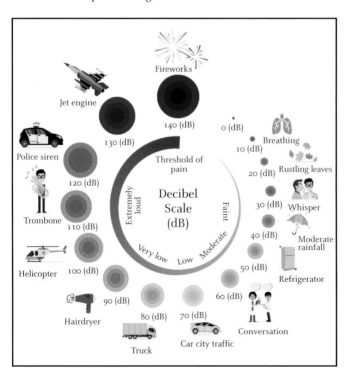

A decibel scale.

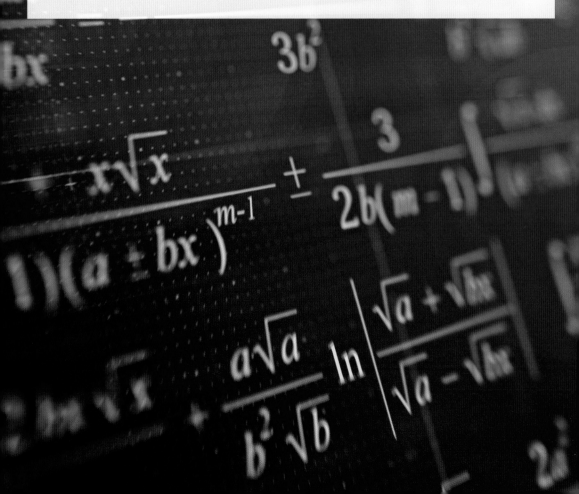

Chapter 11

GEOMETRY GETS CO-ORDINATED

Geometry gets co-ordinated

It's always good to know where things are. One way of solving the problem of locating a point is to employ the system of co-ordinates formulated by French mathematician and philosopher René Descartes.

Descartes proposed that the location of any point in two dimensions can be described by two numbers, one giving the point's horizontal location and the other its vertical location, a system that came to be known as 'Cartesian co-ordinates'.

Before this, mathematics had been split into two distinct branches: numbers and shapes. Descartes' insight meant that, since algebra could be represented as geometry, and vice versa, geometric shapes could be incorporated into a number system. This was the beginning of analytic geometry, a powerful new problem-solving addition to the mathematics toolbox.

Descartes' ideas would lead to the formulation of polar co-ordinates, describing the position of a point using a distance and an angle, by Jacob Bernoulli, and ultimately to one of the most important discoveries in the history of mathematics – calculus.

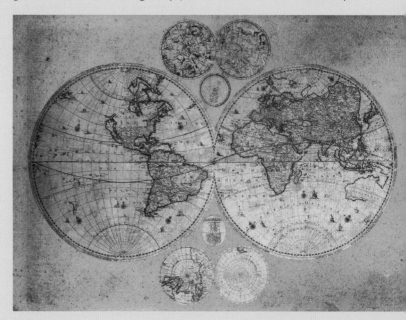

A co-ordinated timeline

1637 — René Descartes publishes *La géométrie*, setting out his ideas for co-ordinate geometry. (Pierre de Fermat had been working on the same ideas but not published them.)

c.1670 — Working independently, Isaac Newton and Gottfried Leibniz build on Descartes' ideas to develop calculus.

1691 — Jacob Bernoulli develops and publishes his work on polar co-ordinates

1692 — Leibniz introduces the first use of the term 'co-ordinates'.

c.1730 — Leonhard Euler, Jakob Hermann and Alexis Clairaut produce general equations for cylinders, cones and surfaces of revolution.

A FUSION OF GEOMETRY AND ALGEBRA

Co-ordinate, or analytic, geometry is a branch of mathematics in which algebraic methods are used to solve problems in geometry. The importance of analytic geometry is that it establishes a relationship between algebraic equations and geometric curves, making it possible to find the solution to an algebra problem by using geometric methods and vice versa. The solutions to an algebra problem can be visualized as a geometric curve and a geometric curve can be expressed as an algebraic equation.

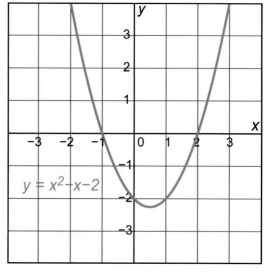

Example of a curve on a graph and its expression as an algebraic quadratic equation.

That numbers and shapes are related was understood by Greek mathematicians. Menaechmus (c.380–c.320BC) proved theorems by using a method that was very close to using co-ordinates. Apollonius of Perga (c.262–c.190BC), known by his contemporaries as the 'Great Geometer', had a great influence on the development of mathematics. His book *Conics* can be seen as a first step towards the development of analytic geometry. He defined a conic as the intersection of a cone and a plane and was able to express properties of the conic as a quadratic equation. He introduced terms we still use today, such as parabola and hyperbola.

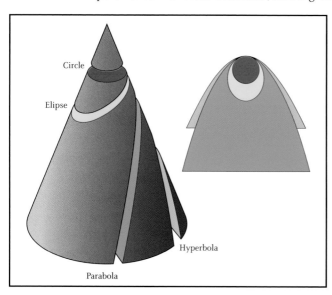

Conic sections.

René Descartes

René Descartes (1596–1650) is perhaps better known today as a philosopher. He believed in taking a rational approach to discover meaning through science and mathematics. He devised a method of deductive reasoning, based on mathematics, that can be applied to all the sciences. Combining mathematics and logic with philosophy to explain how the physical world worked led him to such ideas as the dual nature of mind and body, suggesting that the mind is a mental, or spiritual, substance, whereas the body is a material substance. He also had the curious, though unprovable, notion that our knowledge of things that don't depend on the experience of our senses, but are still generally accepted as true, such as $2 + 2 = 4$ or a cube has six faces, couldn't be relied upon, because God could have made us in such a way that every time we count something we get it wrong! The only thing Descartes felt he could be sure of was that he existed, an idea that, famously, he expressed as '*Je pense donc je suis*' ('I think therefore, I am' – later Latinized as *Cogito ergo sum*).

Frans Hals' portrait of René Descartes.

La géométrie was one of the most influential geometry texts ever written. It was written as an appendix to Descartes' *Discours de la méthode* (Discourse on Method), the book in which he set out his '*Cogito ergo sum*' ideas. In his introduction to *La géométrie*, Descartes stated that: 'Any problem in geometry can easily be reduced to such terms that a knowledge of the length of certain straight lines is sufficient for construction.' He then proceeded to show how problems could be solved through a combination of geometry and algebra.

La géométrie was where he published his idea that two numbers, or co-ordinates, were sufficient to locate the position of any point on a flat surface. It is said, probably fancifully, that the idea came to Descartes while he was watching a fly crawling around on his ceiling. He realized that the path of the fly could be mapped out as a continuous series of points on a plane by reference to two perpendicular number lines, or axes, which cross at a point called the origin. By convention we now name the horizontal axis 'x' and the vertical axis 'y'.

For example, if you wanted to pinpoint the location of the letter F at the start of this sentence you could do so by measuring its distance in from the left-hand side of the page, say x mm, and up from the bottom of the page, y mm, so that the position of the letter on the page is established as x, y.

Pierre de Fermat added a third axis, z, at right angles to the x and y axes, which allowed points to be plotted in three-dimensional space. In fact, it seems likely that Fermat came up with the system before Descartes. In 1636, Fermat was working on a treatise that outlined what we now call co-ordinate or analytic geometry. Unfortunately, although he shared his ideas with other mathematicians such as Blaise Pascal, he never published. Descartes was devising his own system at the same time and published his results in 1637. In the world of science, whoever publishes first takes the credit, which is why we don't have 'Fermatian co-ordinates'!

Death by mathematics

Descartes was not a morning person and would habitually stay in bed (possibly watching flies) until 11am or later. His relaxed routine was disrupted when he travelled to Sweden in 1649 to take up the post of mathematics tutor to Queen Christina. To his horror, he discovered that the queen intended to take her lessons at 5 o'clock every morning. Descartes found it difficult to endure the cold early mornings in Stockholm and within a few months had contracted pneumonia and died, at the age of just 53.

Descartes at the court of Queen Cristina.

Any equation can be represented by plotting its solution set on to the plane. For example, the equation $y = x$ results in a straight line passing through the points (0,0), (1,1), (2,2), (3,3), etc. The equation $y = 4x$ gives a straight line passing through the points (0,0), (1,4), (2,8), (3,12), etc. The solution of more complex equations involving x^2, x^3, and so on result in various types of curve being plotted: $x^2 + y^2 = 6$ produces a circle, for example, while $y^2 - 4x = 3$ results in a parabola.

The points of a co-ordinate change as it rounds the curve and the equation for the curve describes how the value of the co-ordinates change at any point along the curve. A pair of simultaneous equations could now be solved either algebraically or graphically – the values of x and y being the co-ordinates of the points where the lines of the equations intersected.

The fact that any curve, whether a circle, an ellipse, a parabola or hyperbola, can be represented by a quadratic equation is immensely helpful to the scientist as these curves exist in the real world. For example, the path followed by a projectile is a parabola and the path followed by a planet in its orbit is an ellipse.

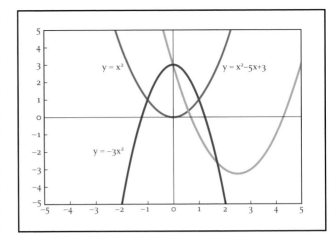

Quadratic equations shown graphically.

MAPS AND MATHEMATICS

Another important use for Cartesian co-ordinates is in cartography. A map isn't of much use if you can't pinpoint your location on it. To solve this problem every map has a grid of numbered lines superimposed on it. The location of any place on the map can be stipulated by just two numbers – one setting it along the east–west axis and the other placing it on the north–south axis.

Maps and mathematics have a long history together. As we have seen, the Egyptian surveyors were skilled

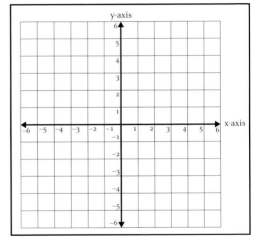

Cartesian coordinates.

at using geometry to measure land boundaries, though there is no evidence that they combined these measurements into maps of large areas. The Greeks, who knew that the world was a sphere and measured its circumference with great accuracy, made many contributions to cartography, including using a grid to indicate the position of a place, a forerunner to the modern latitude and longitude co-ordinate system.

There is one caveat to bear in mind – the Earth isn't flat. For a map of a local area we might, for convenience, pretend that it is, but for a map of the whole world we run into problems. In fact, it is geometrically impossible to use Cartesian co-ordinates to map the Earth on to a flat surface, something that would be proved by Carl Friedrich Gauss's analysis of curved surfaces.

Map of the world using Mercator's projection.

POLAR CO-ORDINATES

Polar co-ordinates are a way of describing the position of something using distance and direction. Rather than specifying a position on a grid by giving x and y co-ordinates, the polar co-ordinates of a point describe its position in terms of a distance from a fixed point (the origin, or pole) and an angle measured from a fixed axis going through the pole. The polar co-ordinates of a point P are therefore given as (r,θ) where r is the distance from the origin O and θ is the angle between Ox and OP. The system was developed by Jacob Bernoulli in the 17th century. A modified three-dimensional version of polar co-ordinates, called spherical co-ordinates, is used by astronomers to pinpoint the positions of celestial objects.

Polar coordinates.

The development of analytic geometry unlocked the way for the discovery of calculus by Newton and Leibniz. It also opened up the possibility of exploring geometries beyond our three-dimensional world. Things that were impossible to visualize in the ordinary sense could now come under the scrutiny of mathematics. It was an idea that would transform mathematics and also physicists' ideas about how the universe works.

Chapter 12

CALCULUS – A SCIENTIFIC REVOLUTION

Calculus – a scientific revolution

A major breakthrough in mathematics was the discovery (or invention) of calculus around the 1670s. Isaac Newton and Gottfried Leibnitz came upon it independently and each vehemently accused the other of plagiarism. Whoever takes the credit, calculus was one of the greatest of all mathematical discoveries. It was developed out of a need to understand continuously changing quantities, a problem that comes up time and again in many fields of science. Calculus was a radical departure from the static geometry of the Greeks; it allowed scientists to begin to make sense of a universe in constant flux. For example, Newton was trying to understand the effect of gravity, which causes falling objects to accelerate. Calculus helped to solve that problem and allowed Newton to establish principles of physics that remained unchallenged until the early 20th century.

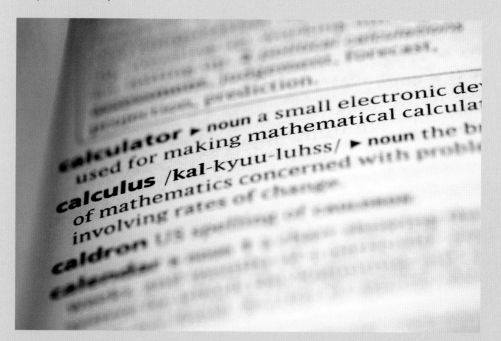

Calculus has become part of the language of mathematics.

Calculus timeline

5th century BC Zeno includes infinitesimals in his paradoxes.

4th century BC Eudoxus of Cnidus uses the method of exhaustion, an early form of calculus.

*c.*1660 Pierre de Fermat publishes work on tangents, maxima and minima.

*c.*1660s Isaac Newton develops his 'method of fluxions' – his version of calculus.

*c.*1670s Gottfried Leibniz develops his version of calculus, the one we use today.

CURVED CALCULATIONS

The origins of calculus stretch back to ancient Greece. Eudoxus of Cnidus (408–355BC) was an eminent mathematician who studied at Plato's Academy. As well as coming up with a theory of proportions that could account for irrational numbers, he also did ground-breaking work on the method of exhaustion, an early form of calculus used to find the area under a curve. It was similar to the method used by Archimedes to determine the value of *pi* – where the area is determined by using successively closer straight-line approximations to the curve. Using his early form of calculus, Eudoxos was able to show that the volume of a cone is one third of the volume of a cylinder large enough to enclose it.

The School of Athens *by Rafael – the Greeks were masters of geometry and laid the groundwork for the development of calculus centuries later.*

The method of exhaustion

This is a method for determining the area of a circle by drawing a series of polygons with an ever-greater number of sides inside the circle and calculating their area. This is done until the space between the perimeter of the polygon and the perimeter of the circle has been almost entirely taken up, or exhausted. The greater the number of sides the polygon has the closer the approximation for the area of the circle.

There is evidence to suggest that mathematicians in India and the Middle East were exploring ideas closely related to calculus some centuries before Newton and Leibniz came along. A version of calculus was invented by Indian mathematician Madhava in the 14th century, for example.

An infinite series is a sum with an infinite number of terms. In the 5th century BC, the Greek philosopher Zeno set out the following paradox, which seemed to show that motion is an illusion. Achilles is running a race with a tortoise. In a spirit of fair play he has given the tortoise a 100-m head start. If Achilles runs 10 times faster than the tortoise, then by the time he reaches the tortoise's starting point, the tortoise will have moved 10m forward. By the time Achilles reaches

that point, the tortoise will have advanced a further metre ... and this will be the case over and over again for ever. By this logic, Achilles will never catch the tortoise! Even though the number of points where Achilles catches up to where the tortoise was last is infinite, the sum of the distances between all those points is finite. This is what is known as a 'convergent series'. An understanding of infinite series would be crucial for the invention of calculus by Newton and Leibniz.

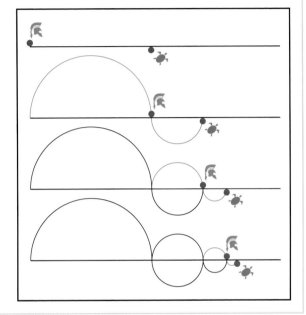

Achilles paradox.

APPROXIMATION

Eudoxos and Archimedes had developed a version of integration – finding the area under a curve. But the other side of the calculus coin, differentiation – finding the gradient of the curve at any one point – was another matter. The analytic geometry developed by Descartes provided a good toolset for tackling the problem.

One way to approach the issue is by approximation. The slope, or gradient, of a straight line is the same wherever you measure it, but the slope of a curve is changing all the time. The change in the curve can be imagined as a series of infinitesimally changing straight lines, each one touching the curve at one point. A straight line touching a curve is known as a 'tangent'. The slope at a particular point on the curve can be approximated by taking the average slope of ever-smaller segments of the curve. As the size of the segment of the curve under consideration approaches zero (that is, an infinitesimal change in *x*), then the calculation of the slope approaches closer and closer to the exact slope at a point. The function that gives the slope of the curve at any point – the slope of the tangent at that point – is called the 'derivative'.

Pierre de Fermat (see pages 80–81) found derivatives for the parabola and hyperbola. He also investigated maxima and minima, the high points and low points of a curve, by considering when the tangent was parallel to the *x*-axis, the point where the derivative is zero. Because of this work, some mathematicians consider Fermat to have been the true 'father of calculus'. Cambridge mathematics professor Isaac Barrow (1630–77), who taught Newton, also described a method of finding derivatives by using tangents.

Newton acknowledged Fermat's contribution when he said: 'I had the hint of this method from Fermat's way of drawing tangents and by applying it to abstract equations ... I made it general.'

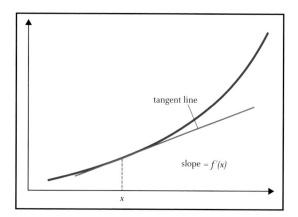

tangent line

slope = $f'(x)$

x

Curve and tangent.

ISAAC NEWTON

Sir Isaac Newton (1642–1727) was one of the greatest scientists of his, or any other era. He is widely considered to have been one of the most influential people in human history. His *Philosophiae naturalis principia mathematica* (the Mathematical Principles of Natural Philosophy), published in 1687 (usually called simply the *Principia*), is ranked among the greatest works of science ever produced. Newton carried out elaborate experiments in optics and founded the science of mechanics. Our understanding of

Take it to the limit

The 'limit' is an important concept in mathematics. It is the value approached as the number of terms in a series tends towards infinity. Taking the limit is a way of dealing with never-ending processes, allowing the mathematician to make sense of the process by taking a series of approximations and determining whether the sequence is approaching a single answer – though it will never actually get there. Limits are an essential part of calculus, making it possible to work with infinitesimals.

gravity and the way objects move through space was pretty much entirely founded on Newton's insights, until Albert Einstein came along at the beginning of the 20th century. Newton also found time to become Master of the Royal Mint, a Member of Parliament and President of the Royal Society, Britain's foremost scientific institution.

Statue of Isaac Newton at the British Library.

As far as mathematics is concerned, Newton's greatest achievement was in developing calculus. As it turned out, he wasn't alone in his thinking. Gottfried Leibniz (1646–1716) also came up with the idea and there was an ugly dispute in which both men claimed credit for the discovery.

Newton didn't call his invention calculus – he called it his 'method of fluxions'. He envisaged a particle tracing out a curve with two moving lines, which were the co-ordinates. A fluxion was the instantaneous rate of change at a particular point on the curve, in other words, the derivative. The changing x and y co-ordinates of the curve were called 'fluents', or flowing quantities. Using his method of fluxions, Newton was able to calculate the slope at any given point on the curve. He wrote about his work on fluxions in October 1666. He didn't publish at the time, but his ideas were seen by many mathematicians and had a major influence on the development of calculus.

ENTER LEIBNIZ

Newton's failure to publish had unfortunate consequences. Over in Germany, Gottfried Leibniz had been working along the same lines. Leibnitz was a philosopher and a diplomat as well as a talented mathematician. Frederick the Great of Prussia once described him as 'a whole academy in himself'. During the 1670s, a few years later than Newton's first discoveries, Leibniz developed a very similar theory of calculus, apparently completely independently. Within a period of about two months he had developed a complete theory of differential and integral calculus. In contrast to Newton, who had made no attempt to make his somewhat unwieldy fluxions accessible to others, Leibnitz took great pains to make his system easily understood and useful.

Statue of Gottfried Leibniz in Leipzig.

Like Newton, Leibniz was a member of the Royal Society in London, so there is a distinct possibility that he was aware of Newton's method of fluxions. Newton certainly got word of Leibnitz's work and in 1676 sent him a 'patent' claim concealed in code, writing: 'I cannot proceed with the explanation of the fluxions now, I have preferred to conceal it thus.' Unlike Newton, however, Leibniz was more than happy to publish his work, and so it was that Europe first heard about calculus from Leibniz in 1684, and not from Newton (who published nothing on the subject until 1693). Leibnitz made no mention of Newton in his publication and when asked about priority answered, 'one man makes one contribution, another man another'. Newton later retorted, 'Second inventors count for nothing.'

When the Royal Society was asked to adjudicate between the rival claims of the two men they gave credit for the first discovery to Newton, and credit for the first publication to Leibniz. However, the Royal Society, by then under the rather biased presidency of Newton himself, later also accused Leibniz of plagiarism, a character assassination from which Leibniz never really recovered. After his death, Newton would boast that he 'broke Leibniz's heart'.

In the end, it was Leibniz's mathematics that triumphed, and his notation and his method of doing calculus, not Newton's clumsier version, is the one that is still used by mathematicians today.

What is calculus?

Calculus gives mathematicians the ability to analyse the rate of change of a quantity over time. It is divided into two categories: differential calculus and integral calculus. Differential calculus deals with the rate of change, such as the acceleration of an object under gravity. Integral calculus is concerned with summing up infinitely small quantities and is used when calculating the area under a curve, giving the net change over time. Calculus is used in many different fields of study, such as the action of waves, the motion of the planets and rates of change in a chemical reaction.

Chapter 13

PICTURE A NUMBER – VISUAL DATA

Picture a number – visual data

The Enlightenment, or Age of Reason, was a flowering of scientific and technological thinking that steadily developed in Europe during the late 17th and early 18th centuries. Among the outcomes of the Enlightenment were the birth of modern science and the beginnings of the Industrial Revolution. From this came a huge new accumulation of data and information. Scientists, engineers and economists needed a new way of handling and comprehending all this data. What was needed was a way of visualizing it.

Today we are very familiar with the idea of visualizing data in the form of graphs and charts, now known as 'infographics', but 250 years ago words and images were two almost entirely distinct ways of conveying information and rarely brought together. William Playfair (1759–1823) changed all that. He was the founder of graphical statistics and inventor of the line graph, pie chart and bar chart. Their impact was enormous at the time and continues to be so today.

The visual language of graphics to display data is one we have grown very familiar with.

Timeline of visual data

1669	Christiaan Huygens uses John Graunt's data on mortality rates to construct a rudimentary graph of the data.
1765 and 1769	Joseph Priestley makes use of graphs in his *Chart of Biography* and his *New Chart of History*.
1786	William Playfair publishes his *Commercial and Political Atlas* with 44 line graphs and a bar chart.
1830s	André-Michel Guerry creates maps of 'moral statistics'.
*c.*1858	Florence Nightingale uses data visualization to draw attention to health conditions during the Crimean War.

THE BIRTH OF INFOGRAPHICS

William Playfair was born in 1759 in Scotland near the city of Dundee, the fourth son of the Reverend James Playfair. His brother John, who was a noted scientist and mathematician of the time, became responsible for William's education when their father died in 1772. William served an apprenticeship with Andrew Meikle, the inventor of the threshing machine, after which he became a draftsman and personal assistant to the great engineer James Watt at the steam engine factory owned by Watt and Matthew Boulton in Birmingham in 1777. It may fairly be said that Playfair's scientific and engineering training came from the very best teachers.

The Storming of the Bastille *by Jean-Pierre Houël.*

In 1789, Playfair, who was living in Paris at the time, took part in the storming of the Bastille prison that heralded the start of the French Revolution. As the Revolution gave way to the Reign of Terror a few years later, he became disenchanted and left Paris.

Playfair was known to be practical and inventive. His knowledge of engineering, based largely on his time spent working with Watt, was first class.

He took out several patents, including one for the first mass-produced silver-plated spoon, and suggested various improvements and modifications for agricultural implements.

An enthusiastic pamphleteer, Playfair often used numerical examples in his political and economic writings. He found that statistical graphics were a huge aid to understanding and believed passionately that graphs were far superior to tables in making large amounts of data comprehensible. In 1786, he published his *Commercial and Political Atlas*, which contained 44 line graphs and a bar chart. This was the first major work of any kind to contain statistical graphs. Playfair's development of the use of statistical graphics between 1786 and 1807 was of such quality and originality that today, nearly two centuries on, very little improvement has been made to his designs.

The first chart in the *Commercial and Political Atlas* illustrates the sum totals of England's imports and exports across the 18th century. The horizontal axis represents one hundred years and is subdivided into decades; from 1760 onwards the subdivisions are made in years. The vertical axis shows money in ten-million-pound increments. The line indicating the monetary value of exports has been coloured red, while that tracing imports is coloured yellow. The area between the two is shaded green when the balance is in England's favour and red when it is not (the latter occurred for a brief time in 1781). In subsequent charts, the same technique is used to illustrate England's trade balance with particular countries and regions. Various other charts show such things as the change in the national debt from 1688 to the time of composition; fees paid for services between 1722 and 1800; expenditure on the navy and the army, and a concluding chart of the price of flour over the previous ten years.

The bar chart that Playfair produced in the *Atlas* illustrated Scotland's imports and exports from 1780–81. It did not include a time element because there was not enough data to produce such a graph. Playfair declared that it was 'much inferior in utility to those that do [include a time element]'. He removed it from subsequent editions of the *Atlas*.

Playfair drew a comparison between his graphical statistics and cartography, which was the reason why he chose to describe his work as an 'atlas'. 'The amount of mercantile transactions in money, and of profit or loss, are capable of being as

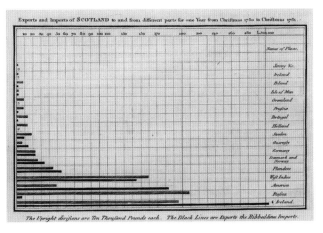

Playfair's graph of Scotland's exports and imports, showing no time element, that he later removed.

easily represented in drawing, as any part of space, or as the face of a country; though, till now, it has not been attempted. Upon that principle these Charts were made,' he wrote. He employed a technique he called 'lineal arithmetic', illustrating it by the example of a merchant who made a pile of the coins he earned each day, lining up the piles so as to make a visual record of his changing revenues day to day. 'Lineal arithmetic is nothing more than those piles ... represented on paper, and on a small scale, in which an inch (suppose) represents the thickness of five million [coins], as in geography it does the breadth of a river...'

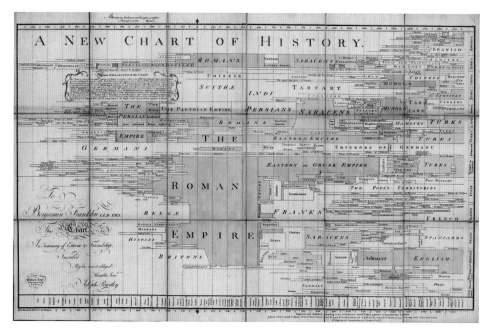

Joseph Priestly's New Chart of History *(1769).*

Playfair realized that any number can be represented as a point on a scale, though he was not quite the first to reach this understanding. Christiaan Huygens had used John Graunt's data on mortality rates to construct a rudimentary graph of the data in a letter to his brother in 1669. Clergyman and scientist Joseph Priestley, the discoverer of oxygen, made use of graphs in his *Chart of Biography* (1765) and in his *New Chart of History* (1769). Priestley, like Playfair, used a geographical analogy, writing of time flowing from beginning to end 'like a river'. He produced a chart of history with a horizontal axis of time and a vertical axis of place. Playfair credited his brother, John, with his inspiration. John had made him keep a daily register of temperature variations, recording them on a divided scale. 'He taught me to know, that, whatever can be expressed in numbers, may be represented by lines,' Playfair wrote.

Playfair's charts made dealing with the explosion of data coming from the work of the Enlightenment's scientists and engineers more manageable. He believed that visualizing numerical data in the form of geometric shapes made it easier to comprehend, easier to remember and easier to pick out significant changes over time. Data, he wrote, should 'speak to the eyes'.

He also emphasized the time that could be saved through data visualization. He thought that as much information could be gleaned in a five-minute perusal of his charts as would be had from days spent poring over tables of figures. 'As knowledge increases amongst mankind, and transactions multiply,' he wrote, 'it becomes more and more desirable to *abbreviate* and *facilitate* the modes of conveying information.'

In 1805, Playfair published *An Inquiry into the Permanent Causes of the Decline and Fall of Powerful and Wealthy Nations*. As its title suggests, the *Inquiry* brought together aspects of Adam Smith's *Wealth of Nations* (1776) and Edward Gibbon's *Decline and Fall of the Roman Empire* (1776–88). Playfair argued that the decline of a nation could be measured and charted and so circumvented. Focusing as he did on national debt and balance of trade, Playfair believed that governments could forestall decline and prolong wealth and power. Economist Henryk Grossman, writing in 1948, described Playfair as the earliest theorist of capital development.

The important thing about Playfair's achievement is that he took Descartes' *x*- and *y*-axis co-ordinates and used them, not to plot mathematical functions, but to represent data. It was an idea that caught on. Soon, data visualization was being used to map things such as the spread of crime and disease in cities. In France, in the 1830s, a lawyer named André-Michel Guerry created maps depicting what he called 'moral statistics'. He was one of the first to employ shading, using darker shades to highlight where crime rates were higher or illiteracy more prevalent, for example. What his maps revealed was that lower levels of education did not necessarily lead to crime, which had been the prevailing wisdom of the time.

By the middle of the 19th century, scientists were even using data visualization to combat epidemics. When London was hit by a cholera outbreak in 1854, physician John Snow mapped out where the incidences of disease had been reported. He noticed a large cluster of cases centred on a water pump on Broad Street. Once

CRIMES CONTRE LES PERSONNES.

Andre Michel Guerry's chart showing incidence of crimes against the person.

the pump had been closed the epidemic began to come under control.

One enthusiastic adopter of visual statistics was British nurse Florence Nightingale (1820–1910). Always a keen student of mathematics as a child, she got her chance to show off her skills during the Crimean War.

Appalled by the filthy conditions of army hospitals and barracks, she persuaded Queen Victoria to let her look into the problem. Together with her friend William Farr, the country's leading statistician, Nightingale embarked on an analysis of army mortality rates. What they discovered astonished them. The leading cause of death among the soldiers wasn't combat – it was disease, and particularly the sort of diseases that could be prevented by good hygiene.

Nightingale realized that the data they had gathered would be best presented visually, 'to affect thro' the Eyes what we fail to convey to the public through their word-proof ears'. What she came up with was the 'polar area chart', an elegant variant on the pie chart. The pie was divided into twelve slices, one for each month of the year, larger or smaller according to the number of deaths, and colour-coded to indicate the causes of death.

Able to see the reality of the situation at a glance, Parliament quickly set up a sanitary commission to improve conditions among the troops. Death rates fell as a result. Florence Nightingale was one of the first people to use data visualization to influence public policy. She would certainly not be the last.

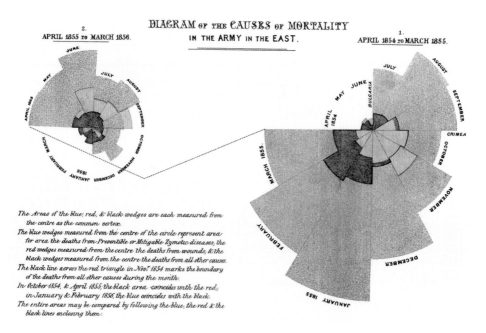

Florence Nightingale's diagram of the causes of mortality in the army based on data she collected during the Crimean War.

Chapter 14

NUMBER THEORY

Number theory

There are patterns to be found everywhere in mathematics. Number theory is about finding and understanding these patterns. It is the oldest branch of pure mathematics, and also the largest. It is all about integers, or whole numbers and their properties, about the relationships between numbers and the patterns and sequences they form, and the various operations we can perform on them. The questions posed by number theory are often easy to state and sometimes surprisingly difficult to answer.

Until the mid-20th century, number theory was considered to be the most 'pure' of the fields of mathematics, with no obvious 'real world' applications. With the advent of digital computers and communications, however, it became clear that number theory did indeed have a role to play in solving real-world problems as it became instrumental in designing powerful encryption schemes used by government and business.

Timeline of number theory

*c.*6th century BC	The Pythagoreans lay the groundwork for number theory.
4th century BC	Eudoxus sets out his theory of proportions.
*c.*4th century BC	Euclid defines number as 'a multitude composed of units'.
3rd century AD	Diophantus writes his *Arithmetica*.
*c.*17th century	Pierre de Fermat revives interest in number theory.
1729	Spurred by correspondence with Christian Goldbach, Leonhard Euler embarks on five decades of fruitful research into number theory.
1801	Carl Friedrich Gauss publishes *Disquisitiones arithmeticae*; it is fundamental in defining number theory and shapes thinking on the subject to the present day.
*c.*1890s	Hermann Minkowski develops a branch of number theory called 'the geometry of numbers'.

MATHEMATICAL BUILDING BLOCKS

The Pythagoreans are considered to have set the foundations of number theory. They were especially interested in sequences of numbers associated with geometric figures. For example, they discovered that the square of any number, n, is equal to the sum of the first n odd numbers. Therefore, if $n = 6$ then $6^2 = 36$ and the first six odd numbers (1 + 3 + 5 + 7 + 9 + 11 + 13) also equal 36. The key thing about Pythagorean number theory was that it involved integers, or whole numbers. To the Pythagoreans, these rational numbers were like cosmic building blocks. As we saw (page 37), this belief put a strain on relationships when a member of the group discovered an irrational number.

Carving in the Cathedral at Chartres, said to be of Pythagorus.

It was Eudoxus, one of the fathers of calculus (page 110), who came up with a theory of proportions that could account for irrational numbers and break through the Pythagorean logjam. As British scientific biographer G. L. Huxley put it, 'Number theory was allowed to advance again ... to the inestimable benefit of all subsequent mathematics.'

In Book VII of *The Elements*, Euclid offered a definition of a number as 'a multitude composed of units' (for Euclid, 2 was the smallest number). *The Elements* also included some important ideas about number theory, particularly concerning the study of prime numbers. Primes are numbers that can only be divided by themselves and 1; they play a crucial role in mathematical thinking. Euclid had two key insights into primes. The first was to prove that there are an infinite number of primes; the list carries on for ever, without end. Secondly, he proved that you can take any non-prime number and break it down into a product of primes. For example, 42 = 2 × 3 × 7, all of which are primes. This idea became known as the fundamental theorem of arithmetic. Any integer greater than 1 is either a prime number, or can be written as a unique product of prime numbers. Like atoms in chemistry, it seems, primes are the building blocks from which all other whole numbers can be assembled.

Further steps forward in number theory were taken in the 3rd century AD by Diophantus of Alexandria. We know very little about Diophantus himself but a series of books he wrote during that time, called the *Arithmetica*, greatly influenced the development of number theory. The number problems Diophantus set became a rich source of ideas for later mathematicians such as Pierre de Fermat. Many of Fermat's discoveries became known by way of the notes he scribbled in the margins of his copy of the *Arithmetica*!

The *Arithmetica* contained 130 equations that came to be known as the Diophantine equations. A Diophantine equation is one where the only permitted variables are positive whole numbers.

Statue of Euclid in the Oxford University Museum of Natural History.

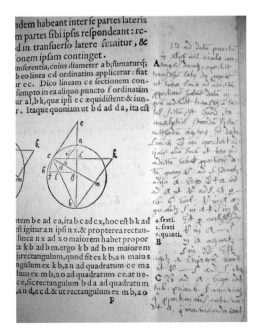

Marginal note by Fermat in a maths book.

For a thousand years following the fall of the Roman Empire, mathematical thinking in Europe was effectively stalled. But this wasn't the case in other parts of the world. In China, India and the Middle East ideas about numbers continued to flow. In India around AD800, for example, the simple and easy-to-use base-ten numbering system – including zero – was being developed and would eventually be adopted by the world's mathematicians.

Mathematics flourished in the Islamic world, too. Ideally situated at a crossroads of world trading routes, Islamic scholars were able to study and learn about the achievements of other civilizations and combine them with their own discoveries.

Although knowledge of the mathematical discoveries of India and the Arab world helped spur the revival in the pursuit of knowledge during the 15th century, in Renaissance Europe the emphasis was more on algebra and geometry. Number theory was a minor pursuit, more fitted to mathematical dilettantes as it had no perceived practical benefits.

FERMAT

Perhaps more than anyone, Pierre de Fermat (1601?–65) can take credit for reviving interest in number theory. Numbers fascinated him and, though he published little, the problems he identified set a course that number theory has followed ever since.

For example, Fermat's little theorem states that a prime must satisfy the condition that if p is a prime number and a any integer, then $a^p - a$ must always be divisible by p.

This simple formula is what lies behind the RSA cryptography system used by banks and other businesses (see page 133 below).

While perusing his copy of the *Arithmetica*, Fermat wrote down one of the most famous, not to say notorious, pronouncements in the history of mathematics. 'To divide a cube into two cubes, a fourth power, or in general any power whatever into two powers of the same denomination above the second is impossible. I have assuredly found an admirable proof of this, but the margin is too narrow to contain it.'

This is what became known as Fermat's Last Theorem, the solution to which would be pursued by mathematicians for more than 300 years until it was finally declared proven in 1995. Andrew Wiles, the British mathematician who made the breakthrough, had to make use of branches of mathematics that were entirely unknown to Fermat, leading some to suggest that Fermat's proof remains to be found, and others to suggest that Fermat never actually had one.

EULER

By the 18th century, mathematics in Europe was far outstripping anything that had been achieved in Greece, India or the Middle East and developments in number theory were a part of that surge. Credit for bringing number theory out of the cold and into the mainstream of mathematics is due in large part to the Swiss mathematician Leonhard Euler (1707–83). Euler was a towering figure in the world of mathematics, making contributions in practically every field of study. Dutch historian of mathematics Dirk Struik described

Euler as 'the most productive mathematician of the eighteenth century – if not of all times'.

To begin with, Euler was as uninterested in number theory as most other mathematicians of the day. It was a correspondence with Christian Goldbach, an enthusiast for number theory, that piqued his interest. At the beginning of December 1729, Goldbach wrote to Euler asking if he was aware of Fermat's observation 'that all numbers $2^{2n} + 1$ are primes?' Euler's response was to demonstrate that Fermat's assertion was wrong by splitting the number $2^{25} + 1$ into the product of 641 and 6,700,417.

Over the course of the next five

Leonhard Euler.

decades, Euler carried out a great deal of research on number theory, much of it tackling Fermat's problems. In 1736, he proved Fermat's little theorem and by the middle of the century he had established Fermat's theorem that primes of a certain form can be uniquely expressed as the sum of two squares.

AMICABLE NUMBERS

Euler also investigated amicable numbers, another topic that had fascinated Fermat. Amicable numbers are pairs of numbers such that the divisors of one add up to the other. The first pair are 220 and 284. 220 can be divided by 1, 2, 4, 5, 10, 11, 20, 22, 44, 55 and 110, which add up to 284; and 284 can be divided by 1, 2, 4, 71 and 142, which adds up to 220. In Euler's time, only three pairs were known, the third of which had been discovered by Fermat. Euler managed to find 58 new pairs!

The second smallest pair of amicable numbers, 1184 and 1210, which had been overlooked by mathematicians of the calibre of Fermat, Euler and others, were discovered by a 16-year-old Italian called Nicolò Paganini in 1866.

There were some problems that even Euler could not solve. He tried and failed to find a solution to Fermat's Last Theorem, though he did make some headway with it.

He gave proofs, or near-proofs, of Fermat's Last Theorem for exponents $n = 3$ and $n = 4$ but despaired of finding a general solution. Euler was fascinated by Goldbach's assertion that any even number greater than 2 can be written as the sum of two primes, and was convinced that the assertion was correct. However, he was unable to prove that it was correct, as can be seen below.

Magic squares

Towards the end of his life, Euler became fascinated by so-called 'magic squares'. These are squares of numbers, in the simplest form from 1 to 9, arranged so that all the rows, columns and diagonals add up to the same total. In 1514, Albrecht Durer produced a 4 × 4 square in which the numbers added up to 34 in an astonishing variety of ways. Euler constructed a modified version of a magic square in which each number or symbol

appears just once in each row or column. He called this a Latin square. Unlike magic squares, Euler's Latin squares have a practical application. They can, for example, be used to work out the order of play in a round robin tournament.

Albrecht Durer's magic square.

Euler's advocacy gave number theory a legitimacy it had previously lacked, which attracted other mathematicians to tackle it, speeding progress and the development of new ideas. In 1770, for instance, Joseph-Louis Lagrange (1736–1813), who was also a renowned astronomer, proved Fermat's assertion that every whole number can be written as the sum of four or fewer squares.

GAUSS

Carl Friedrich Gauss (1777–1855) is sometimes referred to as the 'Prince of Mathematicians' and is ranked as one of the most influential mathematicians who ever lived. He was a child prodigy and many stories are told of his remarkable feats. It is said that, aged only three, he corrected an error in his father's payroll calculations and that by the age of five he was taking care of his father's accounts on a regular basis. At the age of seven, he is reported to have amazed his teachers by adding the numbers from 1 to 100 in seconds. He had spotted that the calculation could be simplified into fifty pairs of numbers each adding up to 101 (1 + 100, 2 + 99, 3 + 98, and so on) giving a total of 5050.

Asteroid hunter

In 1801, the Italian astronomer Giuseppe Piazzi discovered the asteroid Ceres (now called a 'dwarf planet'), causing a stir among astronomers. Unfortunately, it vanished behind the Sun before enough observations had been made to enable its orbit to be calculated with sufficient accuracy to predict where it would reappear. Many astronomers tried to solve the problem of finding the asteroid again, but it was Gauss who succeeded. He did it by employing a technique called the method of least squares, which allows for errors in observations.

Photograph of Ceres taken by NASA's Dawn probe.

Although he made important contributions to almost all aspects of mathematics, Gauss' favourite area of study was always number theory. He declared that 'mathematics is the queen of the sciences, and the theory of numbers is the queen of mathematics'. His *Disquisitiones Arithmeticae*, written at the age of 21 in 1798 and published in 1801, was fundamental in defining number theory and has shaped thinking on the subject to the present day. It was the first textbook to approach algebraic number theory in a systematic fashion. It was also here that Gauss proved that the fundamental theorem of arithmetic, first suggested by Euclid, was true.

Gauss inspired the mathematicians of the 19th century, just as Euler had those of the previous century. It is unfortunate, however, that in later years he appeared to become increasingly arrogant and could be rather dismissive towards those who approached him for advice on mathematical questions. There were even stories of Gauss falsely claiming the ideas of other mathematicians as his own.

NUMBERS AND SPACE-TIME

Towards the end of the 19th century, Hermann Minkowski (1864–1909) developed a branch of number theory called 'the geometry of numbers'. This was a method of solving number theory problems by utilizing a multi-dimensional spatial geometry. It involved

complex concepts such as 'vector space' and 'lattice points'. Minkowski was also one of the young Albert Einstein's teachers. It was Minkowski who, in 1907, realized that Einstein's 1905 special theory of relativity could best be understood in terms of a four-dimensional combination of time and space, a visualization that came to be referred to as Minkowski space-time. Asked about Einstein's theory, Minkowski remarked that it had come as a bit of a surprise, as the young Einstein 'had been a lazy dog ... He never bothered about mathematics at all.'

Hermann Minkowski.

Number theory is concerned with uncovering all the interesting properties that might apply to a natural number. It is often referred to as 'higher arithmetic' and is quite different from elementary arithmetic, which simply uses numbers for the purpose of everyday computing of sums. So, do all numbers have 'interesting properties'?

A well-known mathematical anecdote describes how British number theorist G. H. Hardy went to visit the Indian mathematician Srinivasa Ramanujan (1887–1920) in hospital. Although he had no formal training in mathematics, Ramanujan was famed for his insights into the relationships between numbers. When he arrived, Hardy reported that the taxi in which he had ridden had the license number 1729.

This, he remarked, seemed a rather uninteresting number. Immediately Ramanujan responded that, on the contrary, 1729 was indeed interesting. It was, in fact, the smallest integer that could be expressed as the sum of two cubes in two different ways.

$1729 = 10^3 + 9^3$ and $1729 = 12^3 + 1^3$

There really are interesting numbers everywhere – if you know how to find them!

NUMBER THEORY AND CRYPTOGRAPHY

Mathematician Godfrey ('G.H.') Hardy (1877–1947) once described number theory 'as one of the most obviously useless branches of Pure Mathematics'. Just thirty years after his death in 1947, an algorithm for the encryption of messages was developed using number theory. The Rivest–Shamir–Adleman (RSA) algorithm is one of the most popular and secure public-key encryption methods currently available. The algorithm makes use of the fact that there is no efficient way to factor very large (100–200 digit) numbers. What it does is to take two large, randomly generated prime numbers and multiply them together to generate one very large number. It is probably the most frequently used computer program in the world today, enabling people to make secure payments over the internet, and log in securely to e-mail and other personal services.

PROBLEMS TO BE SOLVED

At the moment, some of the most important problems in mathematics that are still waiting to be solved are found in number theory. They include the Riemann Hypothesis, which dates back to 1859. German mathematician Bernhard Riemann (1826–66) investigated the distribution of the prime numbers, asking the question 'given an integer N, how many prime numbers are there that are smaller than N?' His hypothesis was

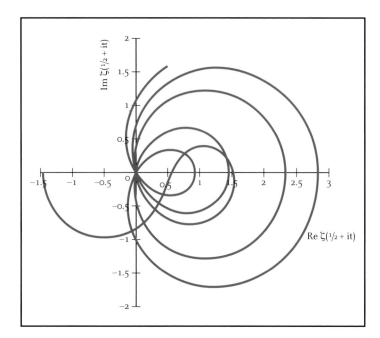

Plot for the Reimann zeta function.

that the distribution of primes was related to something that is now called the Riemann zeta function. If anyone can prove the hypothesis to be true a $1,000,000 prize awaits at the Clay Mathematics Institute in New Hampshire.

One of the oldest unsolved problems in number theory is Goldbach's Conjecture. In 1742, the Prussian mathematician Christian Goldbach (1690–1764) wrote a letter to Leonhard Euler in which he proposed that every integer greater than 2 could be represented as the sum of three primes. Goldbach considered 1 to be a prime number, a convention no longer followed, so today the conjecture is given in the form 'every integer greater than 5 can be represented as the sum of three primes'.

Euler put his own spin on the problem, conjecturing that every even number greater than 2 is the sum of two primes. Euler's version became known as the 'strong' Goldbach's Conjecture.

Computer searches have shown the conjecture to hold for numbers up to 4×10^{18} – that's a 4 followed by 18 zeros. But just because every number we've tried so far works that still doesn't constitute a proof that it will continue to work into infinity. For the problem to be considered solved, a mathematician will have to come up with a way to prove that there are no even numbers for which the conjecture doesn't work. So far, that's another problem that hasn't been solved.

In 2013, the conjecture that all odd numbers greater than or equal to nine are the sum of three odd primes, known as the 'weak' Goldbach conjecture, was proved to be true by mathematician Harald Helfgott (1977–) of the École Normale Supérieure in Paris. He had been working on the problem since 2006.

Chapter 15

THE PROBLEM OF INFINITY

The problem of infinity

The idea of infinity in mathematics is a controversial one – is infinity something that actually exists, or is it just a concept? Is there anything truly infinite in the 'real' world? From earliest times people must have wondered if the world had always existed, or if the star-filled sky above them went on forever. Are these problems mathematics can solve? Dealing with the idea of infinity is perhaps one of the most challenging problems in mathematics and one that mathematicians have struggled with for more than 2,000 years.

Aristotle believed that because infinity doesn't really exist it should have no place in the mathematical universe. In the late 19th century, German mathematician Georg Cantor invented a branch of mathematics dealing with sets – collections of elements that ranged from empty (the equivalent of the number zero) to infinite. Cantor's set theory showed that there could, theoretically at least, be an infinity of infinities.

Does mathematics vanish into infinity?

Timeline of infinity

5th century BC	Zeno uses infinitesimals in his paradoxes.
4th century BC	Aristotle argues that infinities could be potential but not actual.
1st century BC	Lucretius ponders what might happen at the boundary of a finite universe.
12th century AD	Indian mathematician Bhaskara suggests a number divided by zero would be infinity.
*c.*1600	Galileo grapples with the problem of infinity.
1655	John Wallis introduces the infinity symbol: ∞.
*c.*1660s and 1670s	Isaac Newton and Gottfried Leibniz develop calculus using infinitesimals.
1851	Bernard Bolzano's *Paradoxes of the Infinite* is published.
*c.*1874	Georg Cantor introduces the idea of an infinity of infinities.
1924	David Hilbert attempts to explain how infinity works via the medium of an imagined hotel with an infinite number of rooms.

INFINITY IN THE ANCIENT WORLD

Greek mathematicians came up against the concept of infinity early on in their thinking. We've already come across Zeno, the philosopher of the 5th century BC who divided time and distance into an infinity of smaller and smaller segments in his paradoxes (see page 111). There were also the Atomists, who believed that the world was composed of an infinite number of indivisible particles.

Aristotle was not happy with the idea of infinity and introduced a way of thinking that would hold sway for the next 2,000 years. There was, he

(Left) *Does the universe go on forever, stretching out across an infinity of space and time?*

argued, no *actual* infinite, though it was possible to conceive of a *potential* infinite. Mathematicians could, he said, quite easily live without an actual infinite. In his work *Physics* he wrote:

'Our account does not rob the mathematicians of their science, by disproving the actual existence of the infinite ... In point of fact they do not need the infinite and do not use it.'

Even Euclid's proof of an infinity of primes simply said that there are 'more primes than any given finite number', which was, in its way, more of a potential infinity than an actual one.

Even if most mathematicians accepted Aristotle's argument for the potentially infinite, there were still some who came up with some compelling arguments for the actually infinite. In the 1st century BC, Lucretius posed a question in his poem *De rerum*

Lucretius *by Michale Burghers.*

natura ('On the Nature of Things'). Suppose, he asked, that the universe was actually finite. That, then, would mean that there had to be a boundary to it. Suppose you approached that boundary and hurled a stone through it. Where would the stone go? Would it then be beyond the universe? Modern cosmology tells us that it is possible for the universe to be both finite and boundless, but it was an interesting problem to pose and one that would be debated for centuries.

INDIAN INFINITIES

Mathematicians of the Vedic period (around 1500–500BC) in India grappled with some very large numbers. Mantras from before 1,000BC feature powers of ten from a hundred to a trillion, while a 4th-century AD Sanskrit text reports numbering systems leading to a number equivalent to 10^{421} (a one followed by 421 zeroes). As this is hundreds of orders of magnitude greater than the currently estimated 10^{80} atoms in the whole universe, this is perhaps as close to infinity as any ancient mathematicians ever came.

As seen earlier (pages 51–4), Indian mathematicians were also responsible for the introduction of zero into mathematics, using it as a number in its own right rather than simply as a placeholder. In order to do this, they had to make sure that zero was as subject to the rules of arithmetic as any other number. The 7th-century mathematician Brahmagupta established the basic mathematical rules for dealing with zero ($1 + 0 = 1$; $1 - 0 = 1$; and $1 \times 0 = 0$). He also believed that $1 \div 0 = 0$ as well. Almost 500 years later, in the 12th century, another Indian mathematician, Bhaskara II, showed that the answer should be infinity, on the grounds that 1 is be divided into an infinite number of pieces of size zero. However, by this logic every number divided by zero would be infinite, making every number equivalent, since if you reverse the equation then zero times infinity must be equal to every number. The modern view in mathematics is that a number divided by zero is actually 'undefined' (that is, it doesn't make sense).

The largest number smaller than one is...?

What is the largest number that is smaller than 1? Is it 0.999? No, because we can just add another 9 to get a number that is bigger. In fact, we can go on adding 9s into infinity – 0.99999... As there is no number that can be squeezed into the gap between 1 and our infinite string of nines, this results in the seeming paradox that if there is nothing between 1 and 0.99999... on the number line then, effectively 0.99999... equals 1! The only logical conclusion to draw from this is that the largest number smaller than 1 doesn't exist! Indeed, there is no largest number smaller than any number at all.

MEDIEVAL CONUNDRUMS

With a few exceptions, the thinkers of the medieval period were quite happy to leave infinity in the hands of God. St Augustine believed not only that God was infinite, but also that God could think infinite thoughts, and indeed knew all numbers. What madman, as he put it, would say otherwise?

St. Augustine.

There was one interesting paradox involving infinity that the medieval thinkers were aware of. If a line can be divided into infinitely many points then does a circle with a radius of two units have a larger infinity of points around its circumference than a circle with a radius of one unit? Since the circumference of a circle with radius two is twice as long as the circumference of a circle with radius one, then surely the bigger circle should contain a larger infinity of points than the smaller circle. But, since the circles are obviously similar, we can match any point P on the small circle on to any point P' on the large circle, and in the same way map any point Q' on the large circle on to a point Q on the small circle. Seemingly we have two infinities, one larger than the other and yet both the same.

In the early 1600s, the great scientist Galileo Galilei (1564–1642) tried to solve the problem. He proposed that the smaller circle's circumference could be turned into the longer one by adding an infinite number of infinitely small gaps, making them equal in length. He was well aware of the problems caused by this approach: 'These difficulties are real; and they are not the only ones. But let us remember that we are dealing with infinites and indivisibles, both of which transcend our finite understanding, the former on account of their magnitude, the latter because of their smallness.'

In the end, Galileo more or less gave up on the problem of confronting infinity. In what came to be known as Galileo's Paradox, he demonstrated that a

Is the infinite number of points on the circumference of a large circle greater than the infinity of points on a small circle – or are they both the same?

Galileo's tomb in Florence.

one-to-one correspondence could be drawn between all the natural numbers and the squares of all the natural numbers to infinity, a conclusion that suggested that there were just as many square numbers as there were integers, even though it seemed obvious that there were many integers that were not squares. He resolved some of his difficulties by asserting that problems arise only 'when we attempt, with our finite minds, to discuss the infinite, assigning to it those properties which we give to the finite and limited; but this I think is wrong, for we cannot speak of infinite quantities as being the one greater or less than or equal to another'. Galileo concluded that: '...the attributes "equal", "greater" and "less" are not applicable to [the] infinite.'

The infinity symbol: ∞

The symbol ∞ that we use for infinity today (it's called the 'lemniscate'), was first used by English mathematician John Wallis (1616–1703), who used it in *De sectionibus conicis* in 1655 and again in *Arithmetica infinitorum* in 1656. He chose it to represent the fact that one could travel endlessly around its curve.

Abstract infinity symbol.

We saw earlier how Newton and Leibniz developed calculus using the idea of infinitely small quantities, which Newton called fluxions (page 113). Though they undoubtedly produced results, others were wary of these infinitesimal oddities. Irish philosopher George Berkeley asked, 'And what are these fluxions? ... They are neither finite quantities, nor quantities infinitely small, nor yet nothing. May we not call them ghosts of departed quantities?'

Newton believed that space is not just very large but actually infinite. He claimed that such an infinity could be understood, but it could not be conceived. Immanuel Kant, on the other hand, argued that actual infinity couldn't exist precisely because it could not be conceived. In *The Critique of Pure Reason* (1781) he wrote:

'... in order to conceive the world, which fills all space, as a whole, the successive synthesis of the parts of an infinite world would have to be looked upon as completed; that is, an infinite time would have to be looked upon as elapsed, during the enumeration of all coexisting things.'

Hilbert's hotel

German mathematician David Hilbert (1862–1943) tried to explain infinity by imagining a hotel with an infinite number of rooms. Though the hotel is fully booked, room can always be found for another guest. Whenever anyone arrives, the receptionist asks each guest to move to the room with a number one more than their current one, and the new guest now moves into room number 1. One day an infinite number of new tourists arrive all at once, but the receptionist is unperturbed. All that needs to happen is that the current residents all move to the room with a number that is double that of their current one and so an infinite number of empty odd-numbered rooms becomes available for the new arrivals.

David Hilbert.

SET THEORY

One of the most significant developments in the pursuit of a mathematical understanding of infinity came about with the posthumous publication in 1851 of *Paradoxes of the Infinite*, by Bernard Bolzano (1781–1848). In the book, Bolzano argued that the infinite does exist and in the course of his argument introduced the idea of a set, which he defined for the first time:

'I call a set a collection where the order of its parts is irrelevant and where nothing essential is changed if only the order is changed.' Georg Cantor put it succinctly: 'A set is a Many that allows itself to be thought of as a One.'

Why does defining a set solve the problem of making the actual infinite a reality? It works like this. If we think of the integers as a defined set, for example, then there is a single entity, the set of integers, which must be actually infinite. Aristotle, with his notion of a potential infinite was considering the integers from the point of view that we can never conceive of the natural numbers as a whole. However, they are potentially infinite in the sense that given any finite collection of numbers we can always find a larger one. But with the idea of the set each Aristotelean number collection, however large, is simply a subset of the set of integers that must itself be actually infinite.

Russian mathematician Georg Cantor (1845–1918) took Bolzano's ideas and set about putting them on a firm mathematical footing. In 1874, he published an article that marks the birth of 'set theory' – it was soon stirring up controversy. Theoretical physicist and mathematician Henri Poincaré declared set theory a 'disease' from which mathematics might one day recover. In the paper, Cantor considers at least two different kinds of infinity. Before this, orders of infinity did not exist, all infinite collections were considered to be equally infinite.

Cantor considered an infinite series of natural numbers (1, 2, 3, 4, 5 ...), and then an infinite series of multiples of ten (10, 20, 30, 40, 50 ...). Even though the multiples of 10 were clearly a subset of the natural numbers, the two sets could be paired up on a one-to-one basis (1 with 10, 2 with 20, 3 with 30 and so on) demonstrating that they had the same number of elements and were, therefore, the same sizes of infinite sets. Obviously, the same pairing could be applied to any other subset of the natural numbers, such as odd and even numbers. Cantor realized that he could even pair up all the fractions (or rational numbers) with all the whole numbers, thus showing that rational numbers were also the same category of infinity as the natural numbers, even though it seemed somehow intuitively obvious that the number of fractions must surely exceed the number of whole numbers.

It was only when Cantor considered an infinite series of decimal numbers, which includes irrational numbers like π, e and $\sqrt{2}$, that this method broke down. He showed how it was always possible to construct a new decimal number that was missing from the original list, and so proved that the infinity of decimal numbers was in fact bigger than the infinity of natural numbers. The argument can be put that in between each and every rational number there lies an infinity of irrational numbers.

The religious Cantor equated absolute infinity with God and coined a new word, 'transfinite', to distinguish his different levels of infinite numbers from an absolute infinity. Cantor needed a new notation to describe the sizes of infinite sets, and he used the Hebrew letter *aleph* ($\aleph$). He defined $\aleph_0$ (*aleph*-null or *aleph*-nought) as the infinite set of natural numbers; $\aleph_1$ (*aleph*-one) as the set of ordinal numbers and so on. Because of the unique properties of infinite sets, he showed that $\aleph_0 + \aleph_0 = \aleph_0$, and also that $\aleph_0 \times \aleph_0 = \aleph_0$.

A visualisation of algebraic numbers on a complex plane, or z-plane. The complex plane is is a geometric representation of algebraic (complex) numbers. Here, the points becomes smaller as the integer polynomial coefficients become larger.

The possibility of other infinities, for instance an infinity – or even many infinities – between the infinity of the whole numbers and the larger infinity of the decimal numbers was opened up. This idea was known as the continuum hypothesis. Cantor himself believed that there was no such intermediate infinite set, but could not prove it.

He realized that it was actually possible to add and subtract

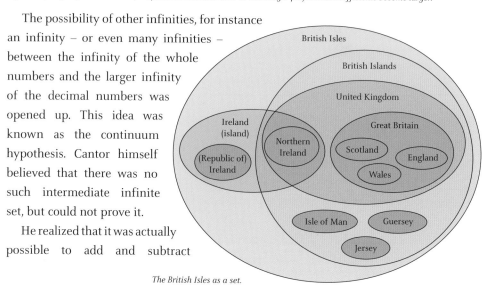

The British Isles as a set.

infinities, and that beyond each infinity there existed another, larger infinity, with yet further infinities beyond that. In fact, Cantor demonstrated, there may be infinitely many sets of infinite numbers – an infinity of infinities – some bigger than others, an idea which, to put it mildly, is somewhat mind-blowing. Responding to Aristotle's potentially infinite, Cantor said that 'in truth the potentially infinite has only a borrowed reality, insofar as a potentially infinite concept always points towards a logically prior actually infinite concept whose existence it depends on'.

Russell's challenge

In 1901, mathematician and philosopher Bertrand Russell issued a challenge to set theory. At first sight, it appeared somewhat innocuous. Imagine listing all the sets that are not members of themselves (the set of odd numbers isn't itself an odd number and so not a member of the set, for example). Suppose we call this big list of sets, A. Russell's question was: 'Is set A a member of itself?'

If we say, yes, A is a member of A, we run into trouble because, by definition, A is a set of sets that are not members of themselves, and so A cannot be a member of itself. But if we say A is not a member of A that can't be true either because we set the criterion for membership of the set as 'a set that is not a member of itself' so according to this rule A had to be in it!

Russell's paradox was a fatal blow for Cantor's version of set theory, but an improved set theory arose to take its place. Because set theory was having a major impact on other areas of mathematics, rather than dismiss it because of paradoxes like Russell's, ways were found to eliminate the paradoxes and keep set theory's main features.

ARE THERE ACTUAL INFINITIES?

Mathematically speaking, there can be no doubt that infinities exist. But is there any quantity out there in the real world that is demonstrably infinite or is it simply an abstract mathematical construct? There are two approaches to the problem – the infinitely big, and the infinitely small.

Is the universe infinite? There is no way we can ever know. It would take an infinite time for light to cross an infinite universe, and we probably won't be around that long! Given when we currently believe the Big Bang took place we can, in theory, see a spherical universe with a radius of about 47 billion light years, which is very big indeed, but not infinite. One cosmology theory suggests that the universe could expand forever, but if it is always, theoretically at least, measurable it'll never be infinitely big.

On a small scale, it has been estimated that there are around 4×10^{80} atoms in the observable universe (give or take a few). That's a four with eighty zeros after it. Again, it's a very big number, but it's not infinite. Quantum physicists think there's a limit to the size of the chunks we can chop space into and though it's very, very small it's still not infinite.

A Planck length is the scale at which our ideas about gravity and space-time cease to make sense. It is 1.6×10^{-35} metres – that's 0.0000000000 00000000000000000000000 16 metres – about 10^{-20} times smaller than a proton. Planck time, which is the time it takes light to travel a Planck length, is 5.4×10^{-44} seconds. These seemingly absolute limits on our understanding look to be as close to the infinitely small as we'll ever get.

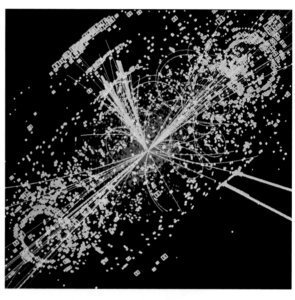

Powerful tools such as the Large Hadron Collider and the Hubble Space Telescope have allowed us to explore the universe on scales from the incredibly small (above) *to the inconceivably vast* (left), *yet acutal infinities remain elusive.*

Chapter 16

TOPOLOGY – THE TRANSFORMATION OF SHAPES

Topology – the transformation of shapes

Topology, sometimes referred to as 'rubber sheet geometry', is the mathematical study of the properties that are preserved through the deformation, twisting and stretching of objects. Cutting or tearing is not allowed. It had its beginnings in 1736 when Leonhard Euler published his solution to the Königsberg bridge problem. Its uses today include such things as the familiar topological map of the London Underground.

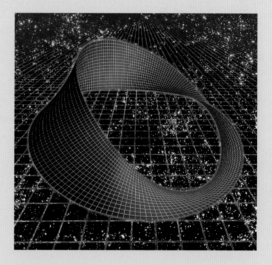

A Möbius strip.

Topological timeline

1736	Leonhard Euler solves the 'bridges of Königsberg' problem.
1750	Euler sets out his formula for a polyhedron in a letter to Christian Goldbach.
1852	The four-colour problem is posed by Francis Guthrie.
1858	August Möbius discovers his famous one-sided loop.

THE BRIDGES OF KÖNIGSBERG

In the 18th century, the city of Königsberg in East Prussia (now the city of Kaliningrad in Russia) was famous for the seven bridges linking the four parts of the city across the River Pregel. The citizens of Königsberg posed a curious question – was it possible to take a walk around the city crossing each bridge exactly once? No one could come up with a solution. Every attempt at making the perfect tour failed, which suggested that it was impossible. But could it be that the right route just hadn't been found yet?

In 1735, the mayor of the nearby city of Danzig brought the problem to the attention of Leonhard Euler, whose work with number theory we have already seen (pages 128–30). Euler felt at first that the problem wasn't one that could be tackled mathematically. The solution, he felt, would be found by reasoning and not by any mathematical principle. Still, he was sufficiently intrigued to tackle it, writing later that it appeared that 'neither geometry, nor algebra, nor even the art of counting was sufficient to solve it'. That being the case, Euler found a novel solution and in doing so laid the foundations for a new branch of mathematics – graph theory.

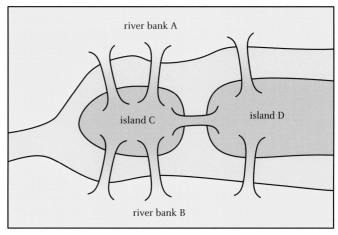

An illustration of the Konigsberg bridge problem.

GRAPH THEORY

Euler's crucial insight was that the actual geographical details of the city were irrelevant to solving the problem. What he did was to reduce the city to an abstract graph, a pictorial representation of the city with the areas of the city as points, or vertices, linked by lines, or edges, representing the bridges. It made no difference if the lines on the graph were of different lengths or whether they were straight or not. Even the choice of route was irrelevant; what was important was the sequence in which the bridges were crossed.

Euler observed that, apart from at the beginning and end of the walk, every time a bridge is crossed to reach another area of the city it must be possible to leave by a different bridge that hadn't previously been crossed. Euler was able to deduce that, because the four areas in the city are touched by an odd number of bridges, it would be impossible to construct a walk that traversed each bridge only once.

In graph theory, each node has a degree, which denotes the number of lines emerging from it. Euler established that it was only possible to trace a non-repeating route if the degree of no more than two of the nodes was an odd number. As all four of the Konigsberg nodes were of odd degree there was no solution to the problem.

Euler published a paper on the solution of the Königsberg bridge problem in 1736. It was entitled, in English translation, 'The solution of a problem relating to the geometry of position.' The title indicates Euler's awareness that he was dealing with a new type of geometry, one in which distance was not relevant.

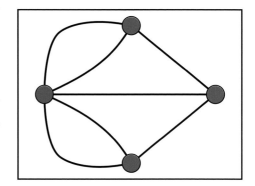

The Konigsberg bridge problem depicted as a graph.

Going underground

The iconic London Underground Tube Map, conceived by Harry Beck in 1933, is an elegant example of a graph. The stations are the vertices and the train lines joining them are the edges. Interchanges are shown where different lines connect. Actual distances and directions in the graph are distorted in the interests of clarity and ease of use.

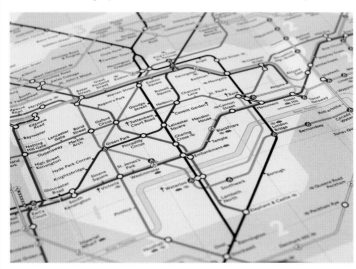

London Underground Tube map.

Euler moved further towards the establishment of a mathematics without measurement in 1750 when he wrote a letter to Christian Goldbach in which he set out his now famous formula for a polyhedron:

$$v - e + f = 2$$

where v is the number of vertices, or corners, of the polyhedron, e is the number of edges and f is the number of faces. For example, for a cube $v = 8$, $e = 12$ and $f = 6$, so $v - e + f = 8 - 12 + 6 = 2$. It is interesting to wonder why this rather simple formula eluded such eminent mathematicians as Archimedes and Descartes, given that both made extensive studies of polyhedra. It has been suggested that before Euler introduced a different way of thinking it was impossible to conceptualize geometry in a way that didn't require measurement.

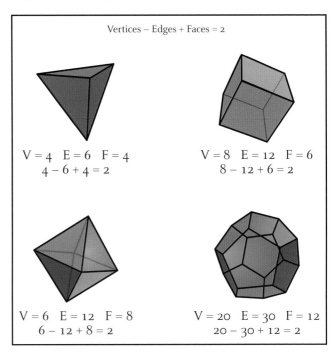

Vertices – Edges + Faces = 2

V = 4 E = 6 F = 4
4 – 6 + 4 = 2

V = 8 E = 12 F = 6
8 – 12 + 6 = 2

V = 6 E = 12 F = 8
6 – 12 + 8 = 2

V = 20 E = 30 F = 12
20 – 30 + 12 = 2

Euler's polyhedron formula.

THE HANDSHAKE LEMMA

The handshake lemma (a 'lemma' is a short theorem used in proving a larger one) is the first theorem of graph theory and grew out of Euler's solution of the Königsberg bridge problem. It states that, in any graph, the number of points with odd degrees must be an even number. You can try it out yourself. Try to draw a graph in which three of the points have an odd degree. It can't be done. The lemma gets its name from the fact that every line in the graph has to begin and end somewhere, just as it takes two people to shake hands. If you add up the degrees of every point in the graph you must get an even number. Adding all the points of an even degree will, obviously, give an even number. So, the total of the points of odd degree has to be an even number too, because subtracting an even number (the points of even degree) from another even number (the degrees of every point) is never going to result in an odd number.

A diagram of the handshake lemma.

The three utilities problem

Imagine three houses, each of which has to be connected up to gas, electricity and water supplies. The catch is that the connections must not cross. Can it be done? The answer is no, it cannot. A graph that can't be drawn in a plane without crossings is called a non-planar graph. It is of course easy to solve the problem by stepping out of the plane into the three-dimensional world, which is what utilities companies do, of course.

INTO THE TREES

A tree is a different sort of graph from the bridges graph or the utilities graph. The Königsberg bridge problem has no defined start or end point. A route through the graph beginning and ending at the same point is called a cycle. A tree graph has no cycles.

Directories on computers are arranged in the form of tree graphs with a root directory having an array of subdirectories branching out from it. Because there are no cycles, the only way to cross from one sub-branch to another is via the root directory.

In a tree graph, each vertex is connected to its neighbour by one path only. The more points there are the more ways there are of joining them up to form a tree graph. There are three ways to draw a five-point tree graph, for example. Tree graphs can be important in organic chemistry where molecules may have an identical number of atoms of hydrogen and carbon but the ways in which the atoms are linked together means that each arrangement has a different chemical property.

The four-colour problem

The four-colour problem was first proposed in 1852 by Francis Guthrie. It basically states that the regions of any map can be coloured using at most four colours so that no two adjacent regions have the same colour. It was finally proved to be true in 1976 by Kenneth Appel and Wolfgang Haken and was the first major theorem to be confirmed using a computer, requiring some 1,200 hours of computer time.

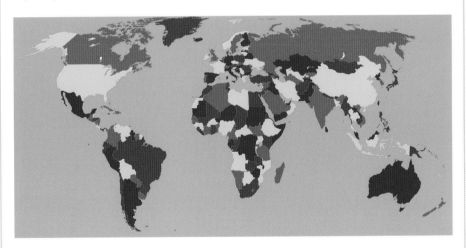

A map of the world showing the four-colour problem solved by Kenneth Appel and Wolfgang Haken.

RUBBER SHEET GEOMETRY

Topology, sometimes referred to as 'rubber sheet' or 'bendy' geometry, is the branch of geometry that deals with shapes and surfaces but doesn't bother with the concerns of regular geometry such as measurements and angles. In that sense, it can be seen how it is related to, and grew from, graph theory. Just as Euler did in solving the bridge problem, topologists turn every shape into a network of nodes and connections that map out the features of a shape that stay the same no matter how much it is distorted.

Of importance to topology are the qualities that do not change when one shape is transformed into another. It is permissible to push and pull, twist and stretch in any direction, all of which are continuous deformations, but not to cut or tear the shape, or to stick one part of it to another part.

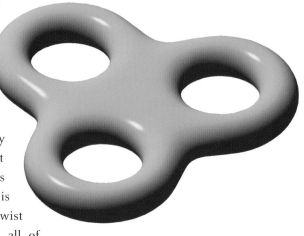

An example of a triple torus – Euler's characteristic of 3.

Cosmic topology

Cosmologists use a lot of topology when they are studying the structure of our universe. The exact shape of our universe, dictated by the amount of matter in it, has very important implications for how it began, how it behaves today, and how it might, at some far distant time in the future, end. Researchers believe the universe could be in the shape of a sphere, a saddle, or even flat. So, you might say that topology has universal applications!

POLYHEDRA

The most basic shapes studied by topologists are polyhedra (the name means 'many faces'). The roots of topology can be traced back to the Greeks. Euclid demonstrated in his *Elements* that there are exactly five regular polyhedra – referred to as the Platonic solids. These are:

The tetrahedron, with four triangular faces.

The cube, with six square faces.

The octahedron, with eight triangular faces.

The dodecahedron, with 12 pentagonal faces.

The icosahedron, with 20 triangular faces.

Each of them obeys Euler's formula $v - e + f = 2$.

Suppose now we take a polyhedron and cut a tunnel through it. Is it still a polyhedron? If we cut a straight-edged hole through a cube then v (number of vertices) now equals 16, e (number of edges) = 32, and f (number of faces) = 16 and Euler's formula fails, as it now gives the answer 0. To make Euler's formula work for all shapes, we have to classify them by the number of holes they contain. Each shape has a value, called the Euler characteristic, calculated by the formula $v - e + f = 2 - 2r$, where r is the number of holes in the object. If there are no holes, as with a regular polyhedron, the Euler characteristic is, as we have seen, 2; if there is one hole, as in a doughnut, it is 0, and if there are two holes, a pretzel shape for instance, it is -2.

Topologists judge different shapes to be the same if one can be pulled or stretched into the other. This can be done providing both shapes share the same Euler characteristic. A typical example of this is the transformation of a doughnut into a coffee cup – both have a surface with a single hole and so are topologically equivalent.

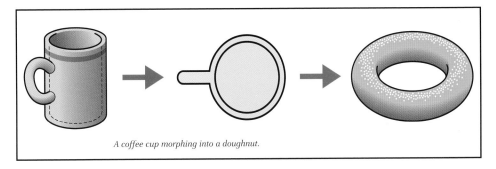

A coffee cup morphing into a doughnut.

MÖBIUS STRIPS AND KLEIN BOTTLES

Ordinarily a surface has two sides. You can't go from one side of a piece of paper to the other without making a hole in it, which is forbidden in topology. However, if you take a strip of paper and give it a half twist before joining the edges together – what do you get? German mathematician August Möbius did just that in the 19th century and discovered that he had a one-sided piece of paper. If you've never tried it for yourself, you really should. Make a Möbius strip then draw a line down the middle with a pencil, you'll find that you can, without lifting the pencil from the paper, draw a line that comes back to where it started.

Another German mathematician, Felix Klein, took things a bit further. He came up with the idea of a one-sided shape – the Klein bottle. He described it as a surface that 'can be visualized by inverting a piece of a rubber tube and letting it pass through itself so that outside and inside meet'. In theory, you can make a Klein bottle by taking two

Möbius strips and joining their boundaries using an ordinary two-sided strip. But don't bother to try as it's impossible to do in three-dimensional space!

The Möbius strip and the Klein bottle are both examples of what topologists call manifolds. The mathematician Leo Moser composed this limerick:

A mathematician named Klein

Thought the Möbius band was divine.

Said he: 'If you glue

The edges of two,

You'll get a weird bottle like mine.'

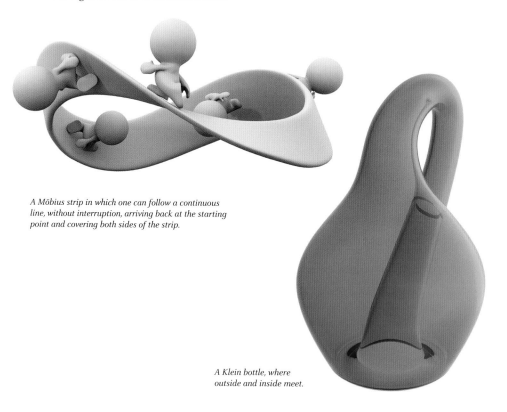

A Möbius strip in which one can follow a continuous line, without interruption, arriving back at the starting point and covering both sides of the strip.

A Klein bottle, where outside and inside meet.

You're a doughnut!

The next time someone tells you that you're a doughnut you're just going to have to accept the fact that, topologically speaking anyway, they're right. A doughnut has a single hole running through it, and so do you. As far as a topologist is concerned, you and the doughnut are equivalent.

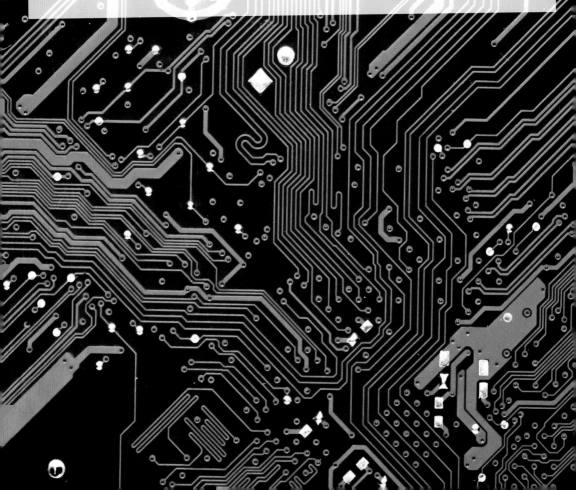

Chapter 17

THE BIRTH OF COMPUTER SCIENCE

The birth of computer science

David Hilbert wanted to know if it was possible to turn any maths problem into an algorithm that would provide a true or false answer without requiring a proof. Was it possible, he wondered, to use the basic logic underpinning the structure of arithmetic to formulate an ultimate theory of mathematics? To solve Hilbert's problem about problems, Alan Turing came up with the idea of a machine, an idealized computer, to demonstrate that it wasn't possible. The theoretical Turing machine, as it came to be called, played a critical role in establishing computer science.

An early desktop computer.

Computing timeline

1642   Blaise Pascal creates the first calculating machine. It uses eight rotating gears and wheels to add and subtract.

1679   Gottfried Leibniz establishes the binary system.

1801   Silk weaver Joseph Marie Jacquard invents a loom that uses punched cards to control weaving patterns.

1822   Charles Babbage published his ideas for the Difference Engine, which can calculate mathematical functions up to six decimal places. The machine has hundreds of gears and weighs two tons.

1931   MIT engineer Vannevar Bush and colleagues build the differential analyser, a computer that solves differential equations.

1935   German inventor Konrad Zuse uses binary notation in his computer designs.

1936   Alan Turing sets out his ideas for what comes to be known as the 'Turing Machine'.

1943   Alan Turing and colleagues design Colossus, a computer employed to decipher Germany's Lorenz encryption during World War II.

1945   John von Neumann lays out a map of computer architecture that is still used today.

WE MUST KNOW! WE WILL KNOW!

David Hilbert was one of the most important and respected mathematicians of the 20th century. His maths genius roamed over many problems and ideas. Many mathematical terms are named after him, including 'Hilbert space' (an infinite-dimensional Euclidean space), 'Hilbert curves', the 'Hilbert classification' and the 'Hilbert inequality', as well as several theorems.

David Hilbert's grave in Göttingen, Germany.

At the 1900 Paris conference of the International Congress of Mathematicians, Hilbert set out the 23 most important open mathematical questions remaining to be answered and so laid the course for 20th-century mathematics. Some of the problems he posed were very precise, while others were vague and subject to interpretation; several of the problems have been solved, at least partially, while a few may never be fully resolved.

Perhaps Hilbert's greatest legacy is his finiteness theorem. He showed that although the number of possible equations was infinite, the number of types of equations was finite and these could be used like building blocks, to produce all the other equations. Hilbert could not actually construct this finite set of equations, he just proved that it must exist – in mathematics this is sometimes referred to as an 'existence proof'. It differs from a constructive, or demonstrative, proof, which directly provides a specific example, or which gives an algorithm for producing an example.

The use of an existence proof rather than a constructive proof was also important in the development of what came to be known as Hilbert space. This extends the geometry of Euclid into spaces with any number of dimensions, even an infinite number. Hilbert space provided the basis for some of the mathematical formulations of quantum mechanics.

A Hilbert space-filling curve is shown here as a continuous fractal space-filling curve.

Hilbert was a great optimist as far as the development of mathematics was concerned. The heading on this section is the inscription on his tombstone. He was convinced that the whole of mathematics could be built on firm logical foundations. His ambition was to find a complete and consistent set of axioms for all of mathematics and this ultimately paved the way for theoretical computer science.

THE TURING MACHINE

In 1935, Alan Turing (1912–54) attended a course on the foundations of mathematics at Cambridge. There, he encountered the three notions that David Hilbert had set forth about the nature of mathematics. Was mathematics complete? Was mathematics consistent? And was mathematics decidable – was there a logical path to find out if a proposition was true or not? Hilbert had assumed that the answer to all three questions would be 'yes'.

(Right) *Sculpture of Alan Turing at Bletchley Park.*

The Turing Machine.

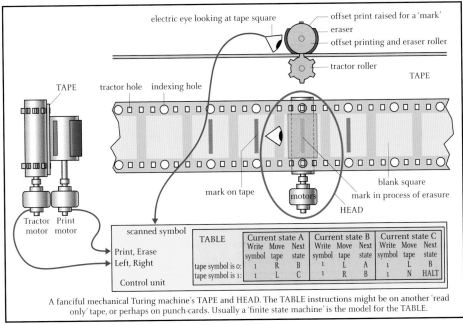

scanned symbol	TABLE	Current state A			Current state B			Current state C		
		Write symbol	Move tape	Next state	Write symbol	Move tape	Next state	Write symbol	Move tape	Next state
tape symbol is 0:		1	R	B	1	L	A	1	L	B
tape symbol is 1:		1	L	C	1	R	B	1	N	HALT

A fanciful mechanical Turing machine's TAPE and HEAD. The TABLE instructions might be on another 'read only' tape, or perhaps on punch-cards. Usually a 'finite state machine' is the model for the TABLE.

In a landmark paper published in 1931, Kurt Gödel (1906–78) had proved that the answer to the first two questions was actually 'no' – arithmetic could not be proven to be both complete and consistent. But the third question remained open. As Hilbert had put it, was there a definite method, a mechanical process, which could be applied to a mathematical statement to determine whether or not it was provable?

Gödel's incompleteness theorem

Kurt Gödel's incompleteness theorem is one of the most influential in mathematics. It can be divided into two parts. First, it states that any theory capable of expressing elementary arithmetic cannot be both consistent and complete: for any consistent, formal theory that proves certain basic arithmetic truths, there is an arithmetical statement that is true, but not provable in the theory. Second, for any formal theory, including basic arithmetical truths and provability, if it includes a statement of its own consistency then it is inconsistent.

In broad terms, what Gödel is saying is that the set of mathematical theorems that are proven is only a subset of the set of theorems that are true. There must be some theorems that are true but which cannot be proved to be true. It effectively put an end to David Hilbert's dream of a complete and consistent basis for all of mathematics.

Kurt Gödel.

In his mind, Turing imagined the construction of an automatic machine that might prove the decidability of any mathematical assertion presented to it. Such a machine would have an infinite memory capacity in the form of an unlimited tape marked off into unit squares, on each of which a symbol could be printed. At any moment, there is one symbol in the machine, called the scanned symbol. The machine can alter the scanned symbol and at the same time its behaviour is partly determined by the symbol. Symbols elsewhere on the tape at the time have no effect on the machine's behaviour, though the tape can be moved back and forth and each symbol could in its turn be scanned and influence the machine's actions.

Any problem might then be reducible to a set of instructions. In essence, what Turing was doing was formulating computer algorithms and the idea of an all-purpose computer. The behaviour of the machine is governed by a table of instructions, an action table – the algorithms – which dictate what is to be done with specific symbols under specific circumstances. The decisive breakthrough was the realization that the action table could be made part of the memory tape. This opened the way for the evolution of the Turing machine into a universal machine that was capable of performing any computable function that was fed into it.

Turing documented his remarkable machine in a landmark paper, the first draft of which he completed in April 1936, 'On Computable Numbers, with an Application to the *Entscheidungsproblem*'. (*Entscheidungsproblem* – German for 'decision problem' refers to Hilbert's notion that mathematics was ultimately decidable.) The paper contained three sections:

1. Defining the ideas of 'computable number' and of 'the computing machine'.
2. Introducing the idea of a 'universal machine'.
3. Employing these ideas to prove that the *Entscheidungsproblem* is unsolvable.

Turing had shown that no machine could solve all mathematical problems. But in doing so, he had laid the groundwork for something of real and lasting importance, the possibility that such a universal machine could be employed to solve many of the practical problems of science and engineering.

Turing earned his doctorate at Princeton University in New Jersey in 1938, when he was 25. Professor of mathematics John von Neumann, arguably the greatest mathematician of his time, offered to make him his research assistant. Von Neumann was impressed by the young Englishman, although his personal letter of recommendation for Turing failed to mention the work on the *Entscheidungsproblem*, which was odd given that von Neumann would later turn wholeheartedly to the pursuit of computing. Then again, as Turing himself wrote to his mother, perhaps solving the *Entscheidungsproblem* wasn't such a big deal when Einstein was just down the hall.

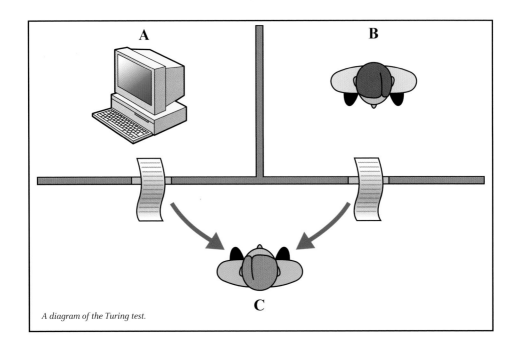
A diagram of the Turing test.

The imitation game

In 1950, Alan Turing addressed the problem of determining whether or not a machine can be said to be intelligent. The test he devised still influences thinking on the subject today. Based on a version of the imitation game, Turing proposed that a questioner interrogates two unseen respondents by means of written questions. One of the respondents is human, the other a computer, and the challenge is to see if the questioner is able to determine which is which. As yet, no artificial intelligence has convincingly passed the test. Although in 2014, a computer chatbot called Eugene Goostman, who had the persona of a 13-year-old boy, convinced 33 per cent of the people who questioned him that he was human.

Turing declined von Neumann's offer and returned to England, where he subsequently joined the codebreakers of project Ultra at Bletchley Park who cracked the German Enigma device with the assistance of Colossus, the first operating computer.

FIRST GENERATION COMPUTING

By the late 1930s, John von Neumann (1903–57) was immersed in the problems of supersonic and turbulent flows of fluids, and by the beginning of the Second World War he had become one of the world's leading experts on shock and detonation waves. He

The binary system

In everyday life, we mostly use the familiar decimal system of counting. The binary system uses just two symbols, 0 and 1, with the numbers ordered in multiples of two, rather than ten. In binary notation 2 would be written as 10 (one 2 and no units) resulting in the hoary mathematical joke that there are 10 kinds of people – those who understand binary and those who don't. Jokes aside, binary is indispensable in computers because every number can be represented by a series of relays that are either on (1) or off (0). The first binary computer was the Z1, invented by German aircraft engineer Konrad Zuse in 1936.

Binary code.

came to realize that computers might help him to solve many of the problems arising from his research interests. By 1944, von Neumann was involved with the Moore School at the University of Pennsylvania, where a group of engineers constructed the first stored-program electronic digital computer, ENIAC (Electronic Numerical Integrator and Calculator). ENIAC used vacuum tubes for circuitry and magnetic drums for memory. It stood nearly 2.4 m (8 ft) high, was 30 m (100 ft) long, 0.9 m (3 ft) deep, occupied a room about 300 square meters in size, and weighed roughly 30 tons. Not really the sort of thing you could have sitting on your desk!

In 1946, von Neumann published a paper that is sometimes referred to as the 'birth certificate of computer science'. He came up with the idea that a computer's data and its instructions should be kept in a single store – the stored-program computer. With the instructions stored in the computer they could be accessed as quickly as needed without first being fed in via paper cards or plugboards.

He also wanted to code the instructions in such a way that they could be modified by other instructions. This was a big step forward because it meant that one program could treat another program as data. Most advances made in writing computer software came about as a result of von Neumann's idea. The way computer components are connected together is called the computer architecture. Von Neumann invented the arrangement of computer components that is still used today in practically every computer. It set out the five principal components of a computer: a unit capable of performing elementary operations of addition; a central processor for executing instructions; a memory for storing data and instructions; and input and output units so machine and humans could communicate with each other. Herman Goldstine, who later directed the computer effort at the Institute for Advanced Study, described von Neumann's report 'as the most important document ever written on computing and computers'.

Von Neumann was an innovator in the field of numerical analysis as well as computer design. Numerical analysis is the technique of solving equations by approximation. It is a technique that can trace its roots back as far as the Babylonians. It developed into a subject in its own right in the 20th century and is an important tool for scientists and engineers who need to model things that change over time, such as stresses and strain on materials, weather forecasting and aerodynamics. Von Neumann's pioneering work in the field still influences the development of the sophisticated computer models used today.

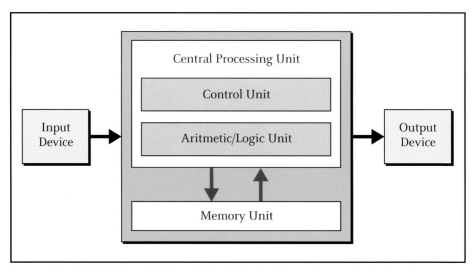

Von Neumann architecture.

Chapter 18

GAME THEORY

Game theory

Strategic thinking plays a vital role in many human activities. Whether it be a pension fund manager assessing where best to invest her money, a player figuring out the best move to make in a chess game, or a general looking for the decisive moment to commit his troops, knowing when and how to act can be of immense importance.

Can mathematics help in making the right decision in competitive games or in conflict situations? Mathematicians like John von Neumann and John Nash believed that it could and developed game theory. Today, game theory plays a central role in economics, diplomacy and sports.

John von Neumann.

Game theory timeline

1928	John von Neumann introduces his theory of parlour games.
1944	John von Neumann and Oskar Morgenstern publish *Theory of Games and Economic Behaviour*.
1950	Merrill Flood and Melvin Dresher pose the Prisoner's Dilemma.
1951	John Nash proposes the Nash equilibrium.

PARLOUR-GAME LOGIC

Game theory is a fairly recent branch of mathematics. Though it has deeper roots, it first came into existence with John von Neumann's paper on his 'theory of parlour games' in 1928. Game theory attempts to determine mathematically and logically the actions that participants should take to ensure the best possible outcomes for themselves in a range of 'game' situations. The game could be an actual game like chess, or something more momentous like warfare. The common factor in each situation is interdependence. This means that the outcome depends not just on what one participant decides to do, but on the decisions made by all the participants. If you decide to sacrifice your rook, will your opponent accept your gambit, or has he or she seen something you missed and is about to do something unexpected?

Game theory only applies to situations where there is a conflict of interest between the participants. It is an important and useful tool in situations where deciding what action will work best for you depends on your expectations of what others will do in their own best interests, and how their actions are determined by what they expect from you. It can't help you decide what to have for lunch.

What is a game?

All games have the following attributes.

1: All participants will act rationally and be bound by the same set of rules.

2: There have to be strategies that the participants can employ to influence the outcome of the game.

3: There has to be a result, or payoff, for the participants' actions.

Chess is a game of strategy.

The mathematics of game theory can be split into two groups: those situations that involve an element of chance, and those that do not. If chance doesn't play a part, analysis can be employed to find a winning strategy that will guarantee victory, if followed correctly. If both players are following the same strategy, the inevitable result will be a draw. A game of chance involves calculating the probabilities of winning or losing, with neither outcome absolutely inevitable.

THE PAYOFF MATRIX

So, what tools does game theory provide to guide our strategic decision making? One example is the payoff matrix.

When you were younger you might have played the game Rock, Paper, Scissors (perhaps you still do!). It's one of the simplest of all strategy games: you count to three and then you and your opponent simultaneously reveal your choice of rock,

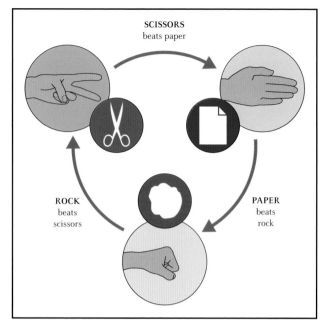

A diagram illustrating the rules of Rock, Paper, Scissors.

paper, or scissors. Rock beats scissors; scissors beats paper; and paper beats rock. Each player makes their choice based on what they think will beat their opponent's choice.

The possible choices, and their outcomes, can be summarized in a payoff matrix. The payoff matrix has three parts: the players involved; the decisions they can make; and the outcomes of those decisions. It illustrates each of the possible strategies available to each player, and every combination of outcomes that could result from their choices. Each game has three choices – rock, paper or scissors – and three possible outcomes – win, lose or draw.

Playoff matrix.

		Player 2		
		Rock	Paper	Scissors
Player 1	Rock	0 / 0	1 / -1	-1 / 1
	Paper	-1 / 1	0 / 0	1 / -1
	Scissors	1 / -1	-1 / 1	0 / 0

A perfect strategy for winning the game would involve knowing exactly what your opponent was going to do and making the best choice to beat them. As that isn't likely to happen, the best strategy available is (as far as possible) to randomize your own choices and so give your opponent no clue as to what you're thinking so they can't formulate a strategy to defeat you. If they are doing the same thing, the most likely long-term outcome over a number of results is a draw.

The Nash equilibrium

In a Nash equilibrium, every person in the game makes the best decision for themselves, based on what they predict the others will do. As each player is going for the 'least worse' option there is no advantage to be gained by changing strategy. The Nash equilibrium encourages co-operation as any change in strategy could result in the player being worse off. It is named after its discoverer, Nobel prize-winning economist John Nash (1928–2015), and is one of the core ideas in game theory. Nash showed that games in which there were a finite number of players and a finite number of choices they could make could reach equilibrium. Roger Myerson of the University of Chicago described Nash's impact on economics as 'comparable to the discovery of the DNA double helix' in biology.

John Nash.

THE ZERO-SUM GAME

Take a look at the payoff matrix for the Rock, Paper, Scissors game and add up the scores in each box. Every one of them sums to zero. In so-called zero-sum games, the interests of the players conflict totally, so that one person's gain is always another's loss. A fundamental rule of game theory is called the 'minimax rule'. This says that there is an optimum strategy that maximizes the minimum gain and minimizes the maximum loss in a two-person zero-sum game. It was proved to be true by John von Neumann in 1928.

THE PRISONER'S DILEMMA

The classic example of game theory at work is known as the prisoner's dilemma. It was first formulated by theorists Merrill Flood and Melvin Dresher in 1950. Imagine two people who have been arrested on suspicion of burglary. They are both guilty of the crime but the police have insufficient evidence to charge them with this. However, the police do have evidence of the lesser crime of trespass. To make the burglary charge stick they will need at least one of the suspects to confess.

Each suspect is held in a separate cell, unable to communicate with the other. Both are made the same offer: if your accomplice stays silent but you confess to burglary and implicate your accomplice, you will go free and your accomplice will go to jail for twenty years. If you both confess, you will both be jailed for five years. If you both refuse to co-operate you will be charged with trespass and go to jail for a year.

What is their best strategy? Following the minimax rule, the best thing to do is to confess. Refusing to co-operate runs the risk of a 20-year sentence if your accomplice confesses and implicates you. Confessing is the only way to minimize the loss and maximize the gain. At worst you will serve a maximum of five years and at best you will go free.

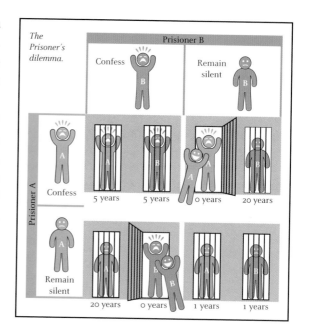

The Prisoner's dilemma.

Burning your boats (or bridges)

Five centuries before game theory was thought of, Spanish conquistador Hernán Cortés was employing strategies that were straight out of the game-theory playbook. When he landed in Mexico with his small force of men, the first thing he did was to very visibly set fire to the ships they had arrived on. This sent a signal to the Aztecs that, their retreat cut off, the Spaniards were going to stand and fight. This had the effect of discouraging the Aztecs from attacking. Cortés had given the impression that he had such confidence in his ability to win any battle that it must surely be folly to fight him. Consequently, the Aztecs retreated. It was Cortés' action that gave us the phrase 'to burn your boats' – meaning that there is no possibility of retreat from a situation.

Hernán Cortés scuttles his fleet.

Chapter 19

CHAOS THEORY

Chaos theory

Is it possible to have a theory that embraces chaos? What sort of theory could encapsulate the seemingly random? Chaos in the mathematical sense doesn't mean disorder. Chaos theory deals with the mathematics of complex systems in which the slightest change in initial conditions can result in drastically different outcomes. Examples of these complex systems are weather systems, turbulence in flowing water and fluctuations in animal populations. Chaos theory explains why the problem of accurate long-term weather forecasting is one we will never solve.

Weather patterns are complex and unpredictable in the long term.

Chaos timeline

1814   Pierre-Simon de Laplace publishes an essay on determinism.

1899   Henri Poincaré publishes his ideas on dynamic instability.

1918   Gaston Julia discovers the first fractals.

1961   Edward Lorenz observes the butterfly effect.

1971   Robert May investigates chaos in animal populations.

1975   Benoit Mandelbrot introduces the term fractal.

THE DETERMINISTIC UNIVERSE

The roots of chaos can be found, ironically, in a belief in determinism. In 1812, Pierre-Simon de Laplace published an essay on the deterministic universe. In it he asserted that were it only possible to know the positions and velocities of all the objects in the universe, and the forces acting upon them, at any one instant it would be possible to calculate their positions and velocities for all future times. Of course, it would never be possible to gather so much data, but the underlying assumption was that we could at least make approximations of how the universe behaves that would be close enough to reality to make no appreciable difference. Chaos theory brought an end to that assumption.

DYNAMICAL INSTABILITY

Around 1900, French mathematician and physicist Henri Poincaré (1854–1912) discovered the phenomenon of dynamical instability. Poincaré was interested in the motions of the planets, thought to be governed by Newton's laws of motion and

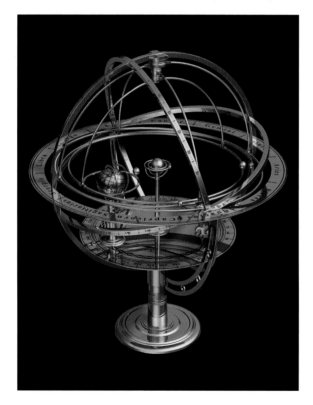

theory of gravity and therefore entirely deterministic. The more precise the measurements made of the positions and motion of celestial bodies, the more accurate would be the predictions of their future positions. Shrinking the uncertainty in the initial conditions would shrink the imprecision in the prediction too. Poincaré discovered that for certain systems this just wasn't the case.

The astronomical systems that did not obey the rule typically consisted of three or more interacting bodies. For these types of systems, Poincaré

An armillary sphere – illustrating how the planets follow fixed paths through space.

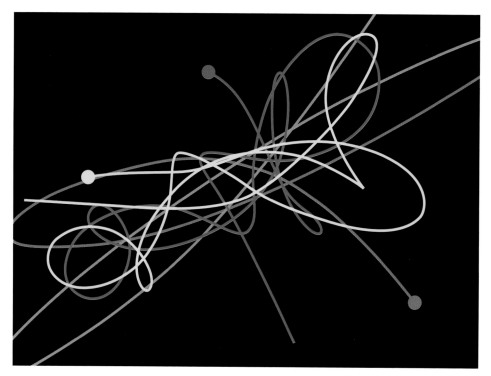

The three-body problem.

showed that a very tiny imprecision in the initial conditions would grow in time at an enormous rate. Poincaré was able to demonstrate that even the tiniest uncertainties in the initial conditions became enormous uncertainties in the final predictions, and this remained so even if the initial uncertainties were shrunk to the smallest imaginable size. This extreme sensitivity of the final outcome to the initial conditions in the systems studied by Poincaré was called 'dynamical instability', and later became simply 'chaos'. Decades would pass before the full implications of his discoveries were appreciated.

THE BUTTERFLY EFFECT

In 1961, meteorologist Edward Lorenz (1917–2008), wrote a basic software program to study a simplified model of the weather. The computer model was based on twelve variables, representing things like temperature and wind speed. Since computer code really is deterministic, Lorenz expected that if he set the same initial values, he would get exactly the same result every time he ran the program.

On this day, however, he repeated a simulation he'd run earlier and was surprised to get a drastically different result. Investigating further he discovered that his second run had included a rounded off value for one of the variables – 0.506 instead of 0.506127. That tiny difference, scarcely more than a ten-thousandth, completely transformed the weather pattern his program produced.

Like Poincaré before him, Lorenz had come to a powerful insight: small changes can have large consequences. The idea came to be known as the 'butterfly effect' after Lorenz pondered the idea that 'the flap of a butterfly's wings in Brazil [could] set off a tornado in Texas'. The butterfly effect, also known as 'sensitive dependence on initial conditions', has profound consequences. Scientists now believe that the weather as a whole is a chaotic system. It would be necessary to take an infinite number of measurements to make long-term weather forecasts with any degree of accuracy. Any uncertainties in the initial measurements, no matter how small, would eventually result in massive inaccuracies in the forecast.

Lorenz's theory that 'the flap of a butterfly's wings in Brazil [could] set off a tornado in Texas' has come to be known as the 'butterfly effect'.

NONLINEARITY

Phase-space maps are a type of graph used by physicists to study the behaviour of physical systems, including chaotic systems. A phase space map might, for example, plot the position and speed of an object. It isn't a graphical representation of what you would see with your eyes if you actually watched the object in motion, since only one variable is a real position.

A phase-space map can show how a system changes and evolves over time. As the system changes, the numbers representing the changes in the phase space also change. The phase space, along with the rules determining how the numbers change, is called a dynamical system. A linear system is one in which a change to a variable always produces the same proportional effect. For example, if you double the change you double the effect, or if you halve the change you halve the effect.

In a nonlinear system, a change to a variable can lead to a change that is not proportional. All chaotic systems, like the weather, are nonlinear. Whereas a linear system can be solved mathematically, a nonlinear system cannot. It really wasn't until the advent of powerful computers that we were able even to approximate the behaviour of nonlinear systems. Despite the best hopes and beliefs of Newton and Laplace, nature is intrinsically nonlinear. In the words of Polish–American nuclear physicist Stanislaw Ulam, 'Using a term like nonlinear science is like referring to the bulk of zoology as the study of non-elephant animals.'

DETERMINEDLY RANDOM

A chaotic system is defined as one that is governed by exact rules, with no element of chance playing a part. In other words, it is a system that is deterministic, but in which random events take place. How is this possible?

In the late 1940s, John von Neumann came up with a deceptively simple way of generating random numbers. You begin with a number, x, multiply x by $(1 - x)$ and multiply the result by 4. Then take the result of that calculation and apply the same formula to it. Once the initial number has been chosen the result is predetermined, but surprising. Suppose we start with $x = 0.2$. The sequence goes 0.64, 0.9216, 0.2890, 0.8219, 0.5855, 0.9707, 0.1137, 0.4031... and so on in a completely random sequence. Today, von Neumann's rule is known as the logistics map and is one of the simplest mathematical illustrations of chaos at work. It was picked up by biologist Robert May in the 1970s, who used it to model changes in animal populations over time. He wrote a paper about his findings called 'Simple Mathematical Models with Very Complicated Dynamics', which seems rather apposite.

ATTRACTORS – STRANGE AND OTHERWISE

The phase-space diagram of a dynamic system can be thought of as possessing attractors. An attractor is a state towards which the system tends to evolve. In the case of a simple system such as a swinging pendulum this would be the rest point of the pendulum.

Edward Lorenz's first weather model involved, as we have seen, a set of twelve variables. He decided to look for complex behaviour using a simpler set of equations, and hit upon the phenomenon of convection – the behaviour of a fluid when it is heated

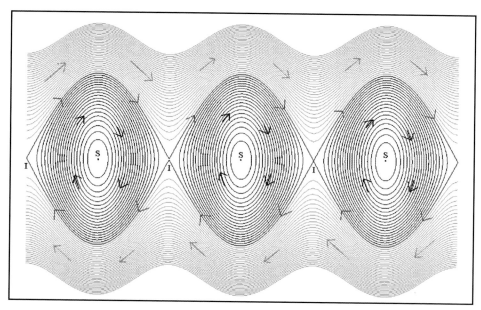

A phases-space map of a pendulum.

from the bottom. He ended up with three equations that were fairly straightforward to solve, but the dynamical system they produced was extremely complicated. When Lorenz plotted the results, he obtained what has come to be called the 'Lorenz attractor'.

The Lorenz attractor is an example of a 'strange attractor'. Strange attractors are unique in that one does not know exactly where on the attractor the system will be. Two points on the attractor that are near each other at one time will be arbitrarily far apart at other times. Unlike regular attractors, strange attractors never repeat. This means that the state

The Lorenz attractor.

of the chaotic system at any one time can never be precisely predicted. No matter how closely the chaotic system phase space is examined, it always exhibits the same degree of complexity and is said to be 'fractal' in nature.

FRACTALS

In 1919, Gaston Julia (1893–1978) carried out a mathematical experiment. He took a simple algorithm and repeated it over and over. Depending on the number he chose for his starting point the results either remained within a limited range or they exploded out. It was only many years later, with the advent of powerful computers, that researchers were able to render Julia's sets graphically and reveal the beauty of their structure.

An example of a Julia set.

Julia's sets were actually among the first fractals. The word was coined by Benoit Mandelbrot (1924–2010). A defining property of fractals is that they remain just as complex no matter how closely you zoom in. They are also self-similar, which means that the structure of the fractal at a small scale is indistinguishable from the structure at a larger scale.

Many things in the real world can be shown to have a fractal nature, such as coastlines and fern leaves. Fractal mathematics has proved to have many practical uses, too – the computer-generated graphics so often used in film-making these days are grounded in fractal generation. Fractal geometry has opened up a path towards understanding complex systems from the occurrence of earthquakes to the workings of financial markets.

Benoit Mandelbrot.

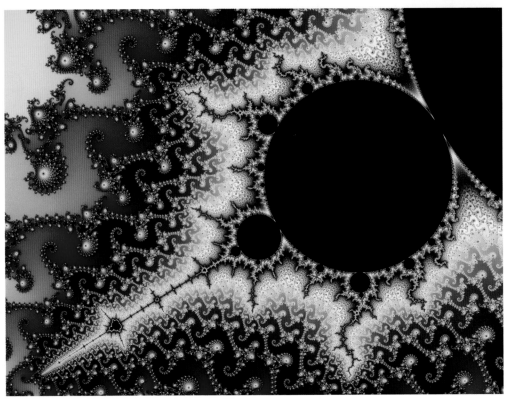

Mandelbrot set.

Chapter 20

CONCLUSION: DOES MATHEMATICS DESCRIBE REALITY?

CONCLUSION: DOES MATHEMATICS DESCRIBE REALITY?

'As far as the laws of mathematics refer to reality, they are not certain, and as far as they are certain, they do not refer to reality.'

Albert Einstein, physicist.

The mathematical physicist and Nobel laureate Eugene Wigner (1902–95) has described mathematics' astonishing ability to describe reality, calling it 'the unreasonable effectiveness of mathematics in the natural sciences'. At the same time researchers are also seeing how ideas from quantum mechanics are influencing thinking about mathematics. So perhaps the ultimate problem to be solved is this: Is reality really mathematical, or is mathematics just something we made up?

Albert Einstein.

'The miracle of the appropriateness of the language of mathematics to the formulation of the laws of physics is a wonderful gift which we neither understand nor deserve.'

Eugene Wigner, physicist and mathematician.

As we saw earlier, the idea that the universe is governed by mathematics is not a new one. Pythagoras had no doubt that the universe was mathematical. The great scientist Galileo Galilei remarked that the universe 'is written in the mathematical language, and the symbols are triangles, circles and other geometrical figures, without whose help it is impossible to comprehend a single word of it; without which one wanders in vain through a dark labyrinth'. But are numbers any more than just something we invented for our convenience; simply a way of balancing the books?

Swedish–American physicist Max Tegmark (1967–) isn't in any doubt. He believes that we have invented a *language* of mathematics that we use to describe the mathematical *structure* of the universe. We can't alter or invent any part of that structure, only attempt to discover it and invent, if need be, a way to describe it. No matter how hard we try, we can't create a reality where the ratio of a circle's circumference to its diameter isn't *pi*.

Tegmark starts with an assumption that Descartes (see page 103) might have taken issue with. He calls it the external reality hypothesis, which states that there is an actual physical reality that exists completely independently of us. We can label the components of this reality with all the fancy names we can dream up, from 'Higgs bosons' to 'pulsars', but none of it needs our consent for its continued existence. According to Tegmark, words used to describe something are just 'optional baggage'. The next step is the mathematical universe hypothesis, which states that the external physical reality we experience is a mathematical structure.

Is there a way, Tegmark asks, to find a description of the external reality that involves no baggage? He believes that the structures of modern mathematics can be defined in a purely abstract way. Mathematical symbols are simply labels; like 'pulsar' they have no intrinsic meaning. It doesn't matter if I write πr^2 or *pi r* squared, it's not going to change what the area of the circle is. It isn't what you call things that matters – what matters is how they relate to each other. The Mathematical Universe Hypothesis implies that the features of the reality we live in come not from the properties of the building blocks of that reality, but from the relationships between these building blocks. Tegmark himself described it as 'this crazy-sounding belief' and it is controversial to say the least.

Max Tegmark (left).

QUANTUM MECHANICS

When physicists began to delve into the realm of quantum mechanics at the beginning of the 20th century they had to turn to mathematics to describe what they found. Probability theory seemed to offer a way forward. Previously, physics had held to the belief that particles, such as an electron, had a definite position in space, but according to quantum mechanics it was only possible to assign a probability of a particle being anywhere. To make things even more awkward, experiments, such as the classic double-slit experiment, had shown that these probabilistic particles could interfere with each other as if they were waves.

Double-slit experiment

The double-slit experiment is one of the most bewildering demonstrations in quantum mechanics. If you send a continuous beam of particles, such as electrons or light photons, through two close-together slits they act like a wave and form a characteristic interference pattern.

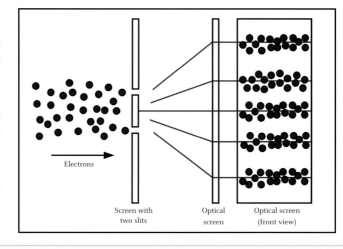

The fun comes when you send through one particle at a time and find that the interference pattern still forms. How does that work? No one knows. As the great physicist Richard Feynman remarked, 'Nobody understands quantum mechanics.'

Double-slit experiment.

Electrons

Screen with two slits

Optical screen

Optical screen (front view)

Classical probability theory was insufficient to the task of explaining this. A solution to the problem was suggested by Austrian physicist Erwin Schrödinger (1887–1961) in 1926. He worked out an equation that determines how probability waves are shaped, how they evolve and collapses when the phenomenon is observed. The Schrödinger equation describes the form of the probability waves (or 'wave functions'), and it specifies how these wave functions are altered by external influences. In effect, the wave function holds all of the measurable information about a particle within it. The wave function can't tell you what you will find when you observe the object, it will only give the probability of which of many possibilities you will see. The Schrödinger equation is said to be as important to quantum mechanics as Newton's laws of motion were for classical mechanics.

The quantum world Schrödinger was describing wasn't something you could picture in your head, like the old 'solar system' model of the atom with electrons spinning around a nucleus. This was a purely mathematical construction. Quantum mechanics described the atomic realm in very precise and rigorous mathematical terms but with outcomes that could only be seen in terms of probabilities and not certainties.

The Copenhagen interpretation of quantum theory, developed in the 1920s, principally by physicists Niels Bohr and Werner Heisenberg, treats the wave function as no more than a tool for predicting the results of observations, and says that physicists shouldn't concern themselves with trying to imagine what 'reality' looks like. Even today, there is still disagreement as to whether the wave function is actually 'real' in any meaningful sense, or whether it is just a mathematical tool that allows us to calculate quantum realm probabilities and therefore has no basis in reality.

Erwin Schrödinger featured on Austrian currency.

WHO COLLAPSED THE WAVE FUNCTION?

Quantum mechanics is one of the best theories science has ever devised to describe how the universe works. Its rock-solid predictive powers have been proved time and again in experiments. Yet the idea that reality is somehow in a fundamental state of uncertainty until observation collapses the wave function was hard to handle, even for the founders of quantum physics. Who collapsed the wave functions to bring the universe into being so we could be here to see it? As Einstein famously taunted, 'Is the Moon not there if no one is looking at it?'

Broadly speaking, there are three alternatives. One is that the wave function is not giving us a complete picture of reality, even though decades of experimentation support the fact that it is. The second alternative is that there is no wave function collapse and every possibility held within the wave function exists in its own separate universe. This is known as the 'many worlds' interpretation.

The third alternative is known as 'objective collapse theory'. This was first proposed in the 1970s by physicists in the United States and Italy. The aim was to make adjustments to Schrödinger's equation so that the wave function evolves naturally from its indeterminate state into a single, well-defined state. To achieve this, they added a couple of extra mathematical terms to the equation: a nonlinear term, which rapidly promotes one state at the expense of others, and a stochastic, or random probability, term, which makes that happen at random.

Richard Feynman.

THE LANGUAGE OF REALITY

The theoretical physicist Richard Feynman said: 'To those who do not know mathematics it is difficult to get across a real feeling as to the beauty, the deepest beauty, of nature ... If you want to learn about nature, to appreciate nature, it is necessary to understand the language that she speaks in.'

Mathematics might not be reality, but it could be that mathematics is the way reality communicates with us. Or is it the case that mathematics fits so well with the way we perceive reality because we invented it do exactly that? Whatever the truth, and it is likely that we will never know, mathematics is probably the best tool humanity ever invented, or discovered, for helping to solve the problem of how reality works.

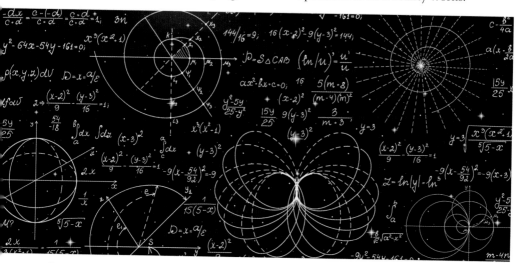

The universe of mathematics.

INDEX